The Institute of Biology's
Studies in Biology no. 159

Mycorrhiza

Richard M. Jackson

Department of Microbiology
University of Surrey

Philip A. Mason

Institute of Terrestrial Ecology
Midlothian

Edward Arnold

First published in Great Britain 1984
by Edward Arnold (Publishers) Ltd
41 Bedford Square, London WC1B 3DQ

Edward Arnold (Australia) Pty Ltd
80 Waverley Road
Caulfield East
Victoria 3145
Australia

Edward Arnold
300 North Charles Street
Baltimore,
Maryland 21201
United States of America

589. 204524
J13m

British Library Cataloguing in Publication Data

Jackson, Richard M.
    Mycorrhiza. – (The Institute of Biology's
    studies in biology, ISSN 0537–9024; no. 159)
    1. Mycorrhiza
    I. Title II. Mason, Philip A. III. Series
    589.2'0452482   QK604

    ISBN 0-7131-2876-3

Printed and bound in Great Britain at
The Camelot Press Ltd, Southampton

General Preface to the Series

Because it is no longer possible for one textbook to cover the whole field of biology while remaining sufficiently up to date, the Institute of Biology proposed this series so that teachers and students can learn about significant developments. The enthusiastic acceptance of 'Studies in Biology' shows that the books are providing authoritative views of biological topics.

The features of the series include the attention given to methods, the selected list of books for further reading and, wherever possible, suggestions for practical work.

Readers' comments will be welcomed by the Institute.

1984
Institute of Biology
20 Queensberry Place
London SW7 2DZ

Preface

Symbiotic associations between fungi and roots are a normal feature of most plants. This fact has been established for many years but is still not widely appreciated. It is no exaggeration to say that a plant's physiology and nutrition cannot be fully understood unless its mycorrhizas are also studied. The past 20 years have seen a great increase in mycorrhizal research world-wide. The aim of this booklet is to provide a simple account of our present knowledge of the subject and to suggest how this knowledge may be of use to improve plant growth and crop yields.

1984
R.M.J.
P.A.M.

Acknowledgements

We would particularly like to thank Dr J. A. Dudderidge, Professor F. T. Last, Dr B. Mosse, Dr G. W. Thomas, Dr M. A. Unsworth and Dr J. H. W. Warcup for their helpful advice and discussion. We are also grateful for the help of numerous colleagues at the University of Surrey and the Institute of Terrestrial Ecology, and to Maureen Hodson who typed the manuscript.

R.M.J.
P.A.M.

Contents

1 What are Mycorrhizas?

1.1 Definition of mycorrhiza

The roots of most healthy plants growing throughout the world, either in a natural or cultivated state, are intimately associated with any one, or sometimes more than one, of many species of fungi. These infected roots are termed *mycorrhizas* (Gr. *mukes*, fungus; *rhiza*, root). The fungal associates are highly specialized parasites which cause little or no damage to their host. The roots of the host plants and the mycorrhizal fungi live together in a balanced or symbiotic relationship in which both partners usually derive benefit from the association (mutualism). The appearance of mycorrhizas varies widely; some look very different from non-mycorrhizal roots, while others can be distinguished from uninfected roots only by microscopic examination.

With few exceptions mycorrhizas have been observed in all plant species of economic importance to man. Controlled experiments have shown that at least some of these plants may not grow or develop normally without mycorrhizas. For example, seedlings of trees, especially pine, without sufficient mycorrhizal roots often fail to survive the first growing season after transplantation from the nursery. In nature also, the mycorrhizal condition is the rule – non-mycorrhizal plants are unusual.

1.2 Classification of mycorrhizas

Mycorrhizas have traditionally been classified into two types, 'ectotrophic' and 'endotrophic'. This classification is based on the arrangement of the fungal vegetative filaments (hyphae) in relation to the host root tissues. Fungi of ectotrophic mycorrhizas enclose the roots in a dense sheath and penetrate into the host cells to a very limited extent. By contrast those of endotrophic mycorrhizas form, at the most, only a loose network of hyphae on the root surface but develop extensively within the root tissues. However, this system for classifying mycorrhizas is not very satisfactory. Although the types within the 'ectotrophic' group are relatively homogenous, three very distinct types of mycorrhizas are included in the 'endotrophic' group – those of the *Ericales* and the *Orchidaceae* and those termed vesicular-arbuscular.

It is now known that functionally vesicular-arbuscular mycorrhizas have much more in common with the 'ectotrophic' than the 'endotrophic' group. Lewis (1973) has suggested a classification of mycorrhizas with four clearly defined groups – sheathing, vesicular-arbuscular, orchidaceous, and ericaceous (subdivided into ericoid and arbutoid). In this book Lewis' system has been adopted to emphasize the fact that there are four rather than two very distinct groups of mycorrhizas. Table 1–1 shows some distinguishing features of these groups.

Table 1–1   Important features of the four major groups of mycorrhizas.

Mycorrhizal group	Sheath	Hartig net	Intercellular hyphae	Intracellular hyphae
Sheathing	+	+	+	−
Vesicular-arbuscular	−	−	+	+
Orchid	−	−	+	+
Ericaceous (subdivided into two)				
Ericoid	−	−	+	+
Arbutoid	+	+	+	+

1.3   Types of relationship

As mentioned previously, the mycorrhizal fungus (the *mycobiont*) usually lives with its plant partner in a balanced, intimate association from which both derive benefit. The mycobiont is heterotrophic and obtains all or most of its carbon and energy requirements directly from its host. At the same time in the sheathing, vesicular-arbuscular and majority of ericaceous mycorrhizas, the fungus keeps its partner supplied with inorganic minerals which it abstracts from the surrounding soil. A characteristic feature of these mycorrhizal relationships, therefore, is the bi-directional movement of nutrients through the fungus. The fungus and host mutually support one another in what is termed a *mutualistic symbiosis*.

In the orchids and some members of the Ericales a very different type of relationship has evolved. These plants possess no chlorophyll for a part or in some cases the whole of their lives. They are able to grow only by obtaining their carbon requirements from the mycobiont. It is probable that the mycobiont also supplies inorganic minerals to these plants, although this is not yet certain. Movement of both carbon and mineral nutrients thus seems to be essentially in one direction only, from the fungus to the plant. This results in the curious situation of the higher plant being parasitic on the fungus which acts as the 'host'.

This discussion of inter-relationships shows that it is not possible to generalize about the mycorrhizal condition. Between the two extremes of mutualism and parasitism described above, there are almost certainly intermediate situations. Also, it is known that a mycorrhizal association may be mutualistic for a period but then become parasitic and sometimes even pathogenic (see below).

It is likely that some plants and their mycorrhizal fungi are examples of two organisms whose evolution has been interdependent, i.e. they have undergone co-evolution.

1.4 Symbiosis, parasitism and saprophytism

Parasitic fungi are those that grow on or in the living tissues of plants or animals. Parasitism may however involve widely differing degrees of association between host and parasite. At one end of the spectrum lies the mutualistic symbiotic relationship between mycorrhizal fungi and their plant partners, in which little or no damage is done to the host tissues. By contrast, many parasitic fungi obtain their nutrient requirements by damaging or destroying their victim. In this situation the host is said to be diseased. The parasite, called a pathogen because it causes disease, secretes large amounts of various enzymes (e.g. hydrolytic pectinases, cellulases and proteinases) into the host causing destruction of the tissues. Such parasitic fungi are therefore well equipped to utilize complex polysaccharides including starch and cellulose. Mycorrhizal fungi generally have a limited capacity to produce extracellular enzymes and probably rely mainly, or even exclusively, on simple carbohydrates such as glucose, fructose and mannose.

Saprophytic fungi either colonize dead plant remains, digesting and absorbing the organic compounds these contain, or they take up soluble organic compounds which have diffused from living or dead organisms. Many are very versatile and able to utilize both simple and complex carbon compounds including carbohydrates, organic acids, sugar alcohols, lipids and alkaloids. The majority of parasitic fungi can live as saprophytes either on the host which they have killed or on other dead organisms; these are called facultative parasites. Others, the so called obligate parasites, can grow only on living host plant tissue.

Many fungi may switch from one type of nutrition to another according to the environmental conditions. For example, two mycorrhizal fungi of certain orchids, *Armillaria mellea* and *Rhizoctonia solani*, can also behave as highly destructive parasites or live saprophytically. They can therefore be considered as either facultative symbionts or facultative parasites. Fungi of vesicular-arbuscular mycorrhizas and the majority of those forming sheathing mycorrhizas do not appear to have an existence independent of their hosts in natural conditions. These fungi are obligate symbionts; the fact that some can be grown in the laboratory on culture media does not necessarily mean that they can also grow saprophytically in the wild.

The concepts of symbiosis and parasitism are described in more detail than is possible here in two other books in this series (Scott, 1969; Deverall, 1981).

2 Sheathing Mycorrhizas

2.1 General description

Sheathing mycorrhizas, also called ectomycorrhizas, are one of the commonest mycorrhizal associations. They occur in the great majority of trees of north temperate countries, e.g. pine, spruce, larch, fir, poplar, willow, lime and oak. In the southern hemisphere many important genera, including *Eucalyptus* and *Nothofagus* (Southern Beech) have sheathing mycorrhizas, as do certain tropical genera. Rather similar mycorrhizas occur in *Arbutus* and related genera in the family Ericaceae (see Chap. 4.)

Many types of sheathing mycorrhizas are formed by fungi that have quite large and conspicuous fruit bodies. Toadstools and earthballs are the commonest examples. This kind of association explains why, for example, fruit bodies of the fly agaric (*Amanita muscaria*) are always found close to certain trees, such as birch and pine. Figure 2–1 shows diagramatically sheathing mycorrhizas of a tree formed on the 'short' feeding roots, and the hyphal cords connecting with fruit bodies.

Trees of all ages, including young seedlings, have sheathing mycorrhizal roots, although the pattern of mycorrhizal development varies during the life of the tree, with some fungi being successful primary colonizers of roots of young seedlings and others coming in at later stages as the tree develops and matures. There is much evidence that mycorrhizas on the root system of a single tree may at any one time be formed by a number of different fungi. This can be easily seen as most sheathing mycorrhizas have a distinctive appearance

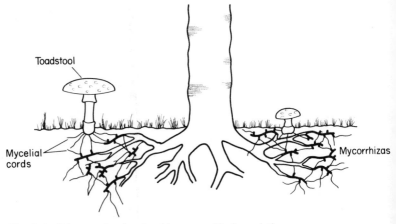

Fig. 2–1 Diagram showing sheathing mycorrhizal association.

quite different from non-mycorrhizal roots and there are usually also obvious differences corresponding to the different species of fungi that form the mycorrhizas.

Simple examination with a hand lens of roots of birch or beech close to the soil surface usually shows that a range of different mycorrhizal forms is present. Roots can either be examined in the field or more easily studied later in the laboratory. It is very important to protect the mycorrhizas from drying up while being transported. In the laboratory the roots should be washed free of all adhering soil particles and examined under a low power microscope ($\times$ 5–50) with top lighting. This will reveal that many beech and birch mycorrhizas, like those of most other species, branch characteristically. However, some of the commonest forms encountered are likely to be unbranched. The best known example of this type, and one that is easily recognizable, is that formed by the fungus *Cenococcum geophilum*. The characteristic dark black, often club-shaped mycorrhizas with stiff, black radiating hyphae are found on the roots of a large variety of trees and shrubs (Fig. 2–2a). The commonest mycorrhizas are pinnate with the main axis giving off branches one at a time, frequently in alternate series; trees such as fir, larch, spruce and also beech possess such mycorrhizas (Fig. 2–2b). These, too, are easy to recognize, as the lateral branches often become progressively smaller towards the apex, giving a christmas tree like appearance. By contrast, some fungi induce a single root to branch prolifically, resulting in a dense aggregate of mycorrhizal tips, sometimes as many as 2000, with a loose to dense sheath. Such *tuberculate* mycorrhizas (Fig. 2–2c) can sometimes even be confused with small puffballs. Species of the genus *Pinus* are unique in possessing simple unbranched mycorrhizas in addition to characteristic 'Y' shaped or forked (bifurcate) ones (Fig. 2–2d) which vary in shape and size according to the species of the fungal partner. The forking varies in complexity, some mycorrhizas being multiple-forked (coralloid). Mycorrhizas of this type arise either directly from the root (sessile) or from a stalk (stipitate).

The unique character of the sheathing mycorrhiza, including its shape, is due partly to the host root, but probably more to the symbiotic fungus. It is the fungus which, after interacting with the root, initiates its distinctive growth pattern. It is now well established that a number of sheathing mycorrhizal fungi are capable of synthesizing growth regulating substances in pure (axenic) culture, and that when cell-free culture solutions of these fungi are added to either excised or intact pine roots, structures resembling coralloid mycorrhizas are formed (see § 7.1.1). Although the transfer of hormones from the fungus to the host plant has not yet been experimentally proven, it seems very likely that the changes which short roots undergo during conversion into mycorrhizas are induced in part by auxin-related substances exuded by the symbiotic fungus.

One of the most striking characters of fresh sheathing mycorrhizas is their wide range of colour. The jet black mycorrhizas formed by *Cenococcum geophilum* have already been mentioned above, while, for example, those formed by the agaric *Paxillus involutus* are distinctly yellow-brown and bruise dark brown just like fruit bodies of this species; by contrast mycorrhizas

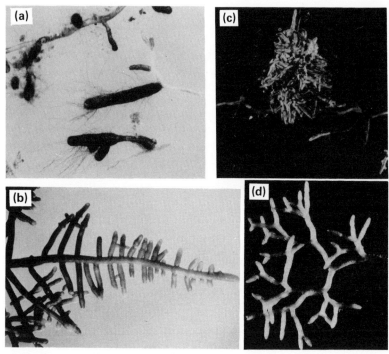

Fig. 2–2 Various forms of mycorrhiza: (**a**) *Cenococcum* mycorrhiza; (**b**) pinnate mycorrhiza (× 3.5); (**c**) tuberculate mycorrhiza (× 1.8); (**d**) dichotomously-branched *Pinus* mycorrhiza (× 3.8). ((**a**) By courtesy of Dr L.V. Fleming.)

formed by *Russula emetica* are pink. However, colour can be misleading as that of fresh mycorrhizas can fade or change quite quickly to another.

Many sheathing mycorrhizas may be enveloped in a dense weft of hyphae while others appear smooth and seem to lack any radiating hyphae or mycelium. In nature, gradations from one extreme to the other may be observed. The attached hyphae may be stiff and bristle-like, or delicate and weft-like (Fig. 2–3a). Simple slide preparations of attached hyphae made by teasing some apart in water or lactophenol and observed at high power (× 400–500) will show their form, whether or not they are septate (i.e. possessing cross walls), and the presence of clamp connections, indicative that the fungus is a basidiomycete. Clamp connections are small loops or protuberances formed at the septa of the hyphae of many basidiomycetes when dicaryotic; i.e. with two nuclei per cell. Characteristic wall ornamentation of hyphae taking the form, for example, of numerous small spines or warts may be visible under the high power (Fig. 2–3b).

In addition to individual hyphae radiating from the sheath there may be hyphal aggregates forming ropes or rhizomorphs, perhaps best termed hyphal cords, originating at the sheath and penetrating far into the soil (Fig. 2–4). These cords may be coloured like the sheath or have other distinctive characters. Whether or not hyphal cords are formed depends mainly on fungal

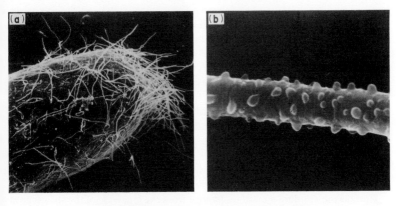

Fig. 2–3   Scanning electron micrographs of (**a**) sheath with radiating hyphae ($\times$ 158); and (**b**) hypha with warts ($\times$ 4500).

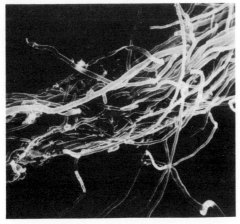

Fig. 2–4   Scanning electron micrograph of hyphal cord ($\times$ 550).

species, for example *Thelephora terrestris* forms cords on Sitka spruce (*Picea sitchensis*) but *Laccaria laccata* on the same tree does not. Various chemical reagents applied to the fungal sheath and attached hyphae produce characteristic colour reactions and can also aid identification. Reagents commonly used include 15% KOH, concentrated NH_4OH, 10% $FeSO_4$, sulfovanillin and Melzer's solution. Chemical colour tests found to be useful for basidiomycetes are described by Rolf Singer (1975).

2.2   Structure

In nature it is mainly the shortest roots of trees that become modified into sheathing mycorrhizas. The first visible sign of mycorrhiza formation is the development of a weft of mycelium over the surface of a short root. The weft gradually thickens over the root surface as it develops into the sheath (also known as the mantle). The sheath often encloses the entire root including its apex so that no direct contact between the younger roots and the soil exists. There is usually complete suppression of root hair development. Thus all nutrients absorbed by the root system must pass through the fungal sheath. Figure 2–5 shows the most important features of sheathing mycorrhizas as seen in transverse section. In addition to its absorptive function the sheath acts as a storage organ, both for carbohydrates and for soil-derived nutrients. Storage of nutrients allows the mycorrhizal fungus to exploit rapidly the optimum period for growth. This is particularly important as trees and shrubs with sheathing mycorrhizas are usually dominant in areas where climate limits growth for considerable periods of the year. The sheath can range in thickness from one to many cells, sometimes being over 100 μm thick. It is also frequently differentiated into two distinct layers. A variety of structures including hairs, cystidia and cords sometimes form on the sheath surface.

Behind the dividing meristem, hyphae from the innerside of the sheath grow between root cortical cells, penetrating to varying depths depending on fungal and tree species. In some mycorrhizas, for example those of beech, the outer cortical cells become radially elongated. This may result in a significant increase in the diameter of the root. This network of hyphae between the cortical cells is termed the Hartig net (after the forest pathologist Robert Hartig). The size and shape of the cells that constitute the Hartig net differ

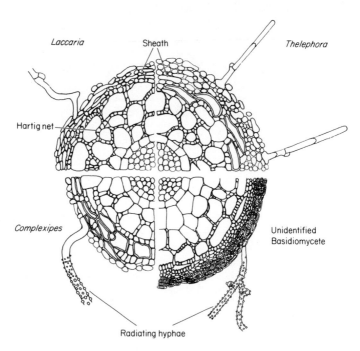

Fig. 2–5 Diagrammatic transverse sections of mycorrhizas of Sitka spruce formed by four different fungi. (By courtesy of Dr G.W. Thomas.)

greatly with different fungi. The form and appearance of the Hartig net and many other microscopical features of the mycorrhiza can best be seen in transverse or longitudinal sections. These can be made quite easily by supporting the mycorrhiza in a piece of carrot or elder pith and using a razor blade to cut sections as thin as possible. The sections are placed in a few drops of 0.05% trypan blue in lactophenol and examined under first a low and then a high power objective. In such sections the Hartig net can be seen to vary from a barely discernible net of hyphae that form bead-like chains between the cortical cells to coarse structures that stand out prominently. The Hartig net has been shown in fact to consist of a three-dimensional branching system which provides a large surface of contact between the fungus and higher plant and is the means by which the two partners exchange nutrients. It is remarkable that once the Hartig net has formed, many cortical cells are completely surrounded by fungal tissue and have no direct contact with other root cells; nevertheless these cortical cells are usually fully active and contain mitochondria, Golgi, endoplasmic reticulum and plastids. The nucleus, however, is not enlarged as it is in mycorrhizas where cell penetration occurs. Depth of penetration of the cortex by the Hartig net ranges from a few cortical cells to the endodermis, but never beyond. Frequently one or more of the outer layers of the cortical cells can be seen to have brown and sometimes granular contents. This is due to the secretion of polyphenols and the affected layer of cells is usually called the tannin layer. There is some evidence that hyphae are adversely affected as they pass between the cells of the tannin layer. However, intracellular penetration of tannin layer cells by hyphae from the Hartig net, in addition to limited penetration of other cortical cells, may occur.

Some mycorrhizas, for example those formed by E-strain fungi in seedling pines, are characterized by a very thin sheath, usually coarse Hartig net and often considerable intracellular penetration. Rather similar mycorrhizas are also formed by genera like *Arbutus* and *Arctostaphylos*, which form the 'arbutoid' group in the Ericaceae (see § 4.1). Mycorrhizas with the above features are termed ectendomycorrhizas.

2.3 Ecology

The certain identification of fungi forming sheathing mycorrhizal associations with a given tree species is usually a difficult and time consuming process. It may involve isolation of the fungus in pure culture (described in § 6.1), often difficult to achieve, followed if possible by its identification and a detailed study of the morphology and microscopic features of the mycorrhiza. It is largely for this reason that, with the exception of a few studies, we know nothing about the frequency and distribution of specific mycorrhizas or how they vary according to season or other changes in the environment. We can infer however, by studying the distribution of fruit bodies, especially those appearing above ground, that in any stand of trees a large number of potential mycorrhiza formers are usually present. It is also important to remember that fruit bodies can give at best only an incomplete picture of the mycorrhizal fungi present. This is because some species only occasionally produce fruit bodies and there are probably many other species that never do. However, from the evidence of fruit bodies some generalizations can safely be made. Conifer woods are characterized chiefly by species of *Suillus, Gomphidius, Lactarius, Russula,* and fungi forming underground or hypogeous fruit bodies, such as species of *Rhizopogon,* while birchwoods, in addition to many species of *Lactarius* and *Russula,* also have *Leccinum, Hebeloma,* and *Amanita* species. Alder has few associated fungi and a correspondingly small range of mycorrhizas occurs on alder roots. On red alder (*Alnus rubra*) growing in Oregon two forms only predominate; one is a rough clavate, dark brown type, probably formed by a host specific hypogeous fungus, *Alpova diplophloeus.* The other, pale brown and glabrous, is thought to be formed by *Lactarius obscuratus,* a fungus usually found only in alder stands in America and central and nothern Europe. Thus some trees form mycorrhizas with a relatively wide range of fungal species while others form symbioses with only a few. A similar degree of variation also occurs with the fungal symbiont. *Suillus granulatus* is probably a symbiont only of pines, while *Paxillus involutus* forms mycorrhizas with a range of trees including pines, spruce, oak, sweet chestnut, hazel and birch. There is now also evidence that there can be marked differences within a fungal species. For example different isolates of *Amanita muscaria* from birch differed in the number of mycorrhizas they formed on birch roots, while an isolate of the same species from pine roots formed no mycorrhizas on birch (Mason, 1975).

The range of mycorrhizas found on tree roots depends on many factors important amongst which is the age of the tree. Young trees appear to form fewer types of mycorrhizas than older trees and they may be fairly distinctive. This has lead to the hypothesis that some mycorrhizal fungi are pioneers and that these are gradually replaced as the tree matures. For example, young spruce and pine seedlings in forest nurseries form mycorrhizas predominantly with what are known as E-strain fungi, one form of which has recently been given the generic name *Complexipes.* These are fungi of as yet uncertain taxonomic position having affinities with ascomycetes. They form large characteristic chlamydospores but no other spores have been seen. In spruce they form typical sheathing mycorrhizas, but in pine they form structures termed ectendomycorrhizas (see § 2.2 p. 00). These fungi are progressively augmented and to a large extent replaced by other fungi following the first year of seedling growth. It seems likely that they are dominant on young seedling roots because they are successful primary root colonizers and are favoured by the physiology of seedling roots and nursery conditions. Some fungi such as *Thelephora terrestris* and *Cenococcum geophilum* may form mycorrhizas both in the nursery and with more mature trees in the forest. Others, however, seem to form part of a succession, being dominant on either seedlings, semi-mature or mature trees.

The roots of young trees growing on adverse sites, including coal and kaolin spoils, in the United States have very few types of mycorrhizas. A type frequently dominant is thick, bifurcate to coralloid and yellow to gold in colour, and is formed by *Pisolithus tinctorius,* a gasteromycete related to puffballs and stinkhorns. The ecological adaptability of this fungus is due to its ability to tolerate high soil temperatures and extreme soil pH. Drained peat bogs are environments that favour only relatively few sheathing mycorrhizal fungi. In these are found, for instance, *Cortinarius pholideus*, *Lactarius helvus* and *Russula venosa*.

2.4 Fungi forming sheathing mycorrhizas

It is not surprising that the best known mycorrhizal fungi are those forming conspicuous fruit bodies above ground. The majority of these are Hymenomycetes, species in which basidia and basidiospores form in a well-developed layer, the hymenium. Several fungi of this type have already been mentioned in § 2.1. As our knowledge increases it is becoming very apparent that fungi in several major taxonomic groups other than the Hymenomycetes frequently form sheathing mycorrhizas. Examples of some of these fungi and their taxonomic position are shown in Table 2–1.

Table 2–1   Sheathing mycorrhizal fungi.

Subdivision	Class	Genus	Fruit body
Zygomycotina	Zygomycetes	*Endogone*	Underground – small, truffle-like
Ascomycotina	Hemiascomycetes	*Elaphomyces* ? *Cenococcum*	Underground – false truffle
	Discomycetes	*Tuber*	Underground – truffle
		Sepultaria	Ground – small, elfin cup
	?	*Complexipes*	Not known – chlamydospores
Deuteromycotina	Hyphomycetes	*Chloridium*	Absent – asexual spores
Basidiomycotina	Hymenomycetes	*Amanita*	Toadstools (gills)
		Hebeloma	" "
		Lactarius	" "
		Russula	" "
		Paxillus	" "
		Boletus	" (pores)
		Suillus	" "
		Thelephora	Bracket (rough surface)
	Gasteromycetes	*Rhizopogon*	Underground – truffle-like
		Scleroderma ⎱ *Pisolithus* ⎰	Ground – earth ball

There is little doubt that the majority of sheathing mycorrhizal fungi are ecologically-obligate symbionts, i.e. in natural conditions they can either not grow at all or grow only to a very limited extent unless they form a symbiotic association with a host. Some of these fungi have so far proved impossible to isolate in pure culture and are therefore analogous to the vesicular-arbuscular fungi, while others that have been isolated grow very slowly. Yet others are quite easy to isolate and grow well in pure culture. The latter group are often characterized by a wide host range and may even be capable of some saprophytic growth under natural conditions.

3 Vesicular-arbuscular Mycorrhizas

3.1 General description

Vesicular-arbuscular (V–A) mycorrhizas occur very commonly in a wide range of plants, including pteridophytes, gymnosperms, angiosperms, and occasionally bryophytes. There are only a few angiosperm families in which V–A mycorrhizas are absent or rare, and which do not appear to form any other kinds of mycorrhizas; examples are Cruciferae, Chenopodiaceae and Resedaceae. V–A mycorrhizas are characterized by a coarse branching mycelium growing through the soil, connected at a few points to mycelium in the roots of host plants. The mycelium in the roots grows within and between cortical cells and usually gives rise to two types of structure. Arbuscules, formed in cortical cells on branches of the hyphae are composed of repeatedly-branching hyphae, the branches progressively diminishing in size (arbuscule means a dwarf tree). The other structures, often developing later than arbuscules, are inter- or intracellular spherical to oval vesicles. Spores, often very large, are formed on the external mycelium. Differences in spore structure are used in the identification of genera and species of V–A fungi. The main features of V–A mycorrhizas are shown diagrammatically in Fig. 3–1.

The fungi forming V–A mycorrhizas belong to a few genera in the class Zygomycetes; until quite recently they have usually all been referred to as species of the genus *Endogone*. Most species show little specificity, being able to form mycorrhizal associations with a wide range of host plants. All attempts to culture V–A fungi have so far failed.

Interest in V–A mycorrhizas has intensified during the past 20 years as it has been realized that the great majority of crop plants normally form V–A mycorrhizas and that these may profoundly affect the crops' nutrition and possibly other aspects of their growth.

3.2 Structure

V–A mycorrhizal infections usually have little or no effect on the morphology of the roots, but can result in a change of colour. For example, infected onion roots may be distinctly yellow. An increase in the size and in the amount of branching of infected roots have also occasionally been noticed. Root hairs are not suppressed in infected roots and in some hosts infection may actually occur through root hairs.

Spores are formed by V–A mycorrhizal fungi in the soil, either singly and borne directly on the hyphae, or in groups within a fruiting structure termed a sporocarp. *Glomus mosseae* is an example of a sporocarp-forming species. The

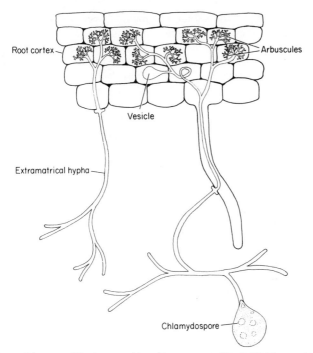

Root cortex

Arbuscules

Vesicle

Extramatrical hypha

Chlamydospore

Fig. 3–1 Diagram of V–A mycorrhiza. (By courtesy of Dr G.W. Thomas.)

sporocarps of this species consist of a more or less spherical cluster of asexual chlamydospores enclosed within a loose wall composed of branching and anastomosing hyphae: the whole fruiting structure may have a diameter of up to about 1.0 mm. Young spores have a thin hyaline wall, but as they mature a much thicker endospore wall is laid down inside the original spore wall which remains as a thin transparent envelope. Mature spores have a distinctive funnel-shaped attachment and range up to more than 300 μm in diameter. The spores of this and some other V–A mycorrhizal fungi are amongst the largest single-celled fungal spores known (Fig. 3–2a). In place of the asexual chlamydospores of species like *G. mosseae*, members of the genus *Endogone* form sexual zygospores. Zygospores typically develop following the fusion of two gametangia, and when mature may be up to 800 μm or more in diameter and bear a swollen suspensor. Zygosporic species may also form distinctive vesicles singly or in groups in the soil. *Endogone* species do not form V–A mycorrhizas although they are thought to be closely related to species of genera such as *Glomus* that form V–A mycorrhizas but not zygospores.

The external or extramatrical mycelium of V–A fungi grows more or less extensively through the soil and is connected with roots only via penetration points on the roots surface. It consists of two fairly distinct types of hyphae, although there are also intermediate forms. Coarse thick hyphae form a network that persists in the soil for a considerable period. These hyphae are generally aseptate; they have characteristic angular projections and may also have dense protoplasmic contents with conspicuous oil droplets. The second type of hyphae arise from the coarse thick-walled hyphae and are much finer, thin-walled, and usually hyaline. They are initially in protoplasmic connection with the thick-walled hyphae, but become septate, lose their contents and disintegrate.

The internal root phase of V–A fungi can easily be studied microscopically without sectioning the roots. Portions of roots, or even whole root systems of seedlings, are carefully washed in water to free them from soil. They are then placed for 30–60 minutes in 10% KOH at 90°C, to increase their transparency, and subsequently stained in 0.05% trypan blue in lactophenol. Following the staining, the roots are mounted in lactophenol on microscope slides for examination.

Roots are normally penetrated by V–A fungi at rather few points either through epidermal cells or occasionally root hairs. Each penetration results in limited colonization of root tissue, possibly not more than about 0.5 cm of root length per penetration. When an infection has been established, runner hyphae sometimes grow over the root surface for some distance from the point of original infection and then produce one or more further infections. Fungal spread within and between the host cells leads to the formation of the typical mycorrhizal structures, arbuscules and vesicles, chiefly within the middle and inner cortex. As in sheathing mycorrhizas, no colonization of tissues within the endodermis takes place. A few fungal species, probably closely related to V–A mycorrhiza formers, may develop a sheath on the roots of some plants. An example is *Endogone eucalypti*, forming a sheathing mycorrhiza on the roots of seedlings of certain species of *Eucalyptus* in Australia.

Arbuscules develop when a hyphal tip or branch enters a host cell and starts to branch dichotomously, such that the finest branches cannot be resolved with a light microscope (Fig. 3–2b). As a result, the arbuscule provides an extremely large area of contact between the fungus and the cell cytoplasm. The cytoplasm of the arbuscules, like that of the external and internal hyphae, contains many small vacuoles, within the majority of these vacuoles are small spherical granules about 0.1–0.2 μm in diameter. These granules have been shown by histochemical staining reactions to consist of polyphosphate and are believed to be concerned with the vacuoles in the transport of phosphate by cytoplasmic streaming through hyphae from the soil to the arbuscules. The arbuscule branches are almost certainly the main sites of nutrient exchange between fungus and plant. Within the host cell, the nucleus and nucleoli are enlarged and the volume of cytoplasm is increased. After a rather limited period of activity, arbuscules lyse, probably with release of further nutrients to the plant.

Vesicles, swellings along or at the tips of hyphae, develop mostly between primary cortical cells. Initially vesicle walls are thin but they thicken and lipids are deposited in their cytoplasm. It is assumed that vesicles serve both as storage structures and sometimes as survival propagules when the infected root dies and disintegrates.

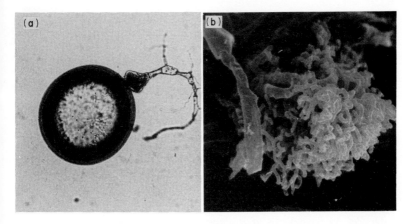

Fig. 3–2 (**a**) Micrograph of a chlamydospore of a V–A fungus (× 165); (**b**) scanning electron micrograph of an arbuscle (× 1430). ((**b**) By courtesy of Dr C.M. Hepper.)

3.3 Ecology

Since a method for separating spores of V–A fungi from soil, based on wet sieving and decanting, was first developed by Gerdemann in 1955, it has been used widely to study some aspects of the ecology of these fungi. A version of this method that gives good results is as follows. A composite soil sample is thoroughly mixed, a 50 g portion then shaken with a litre of water and allowed to sediment for 10–20 seconds in a litre measuring cylinder. This removes mineral and heavier organic particles from the suspension, which is now passed through a 700 μm mesh sieve. Fine roots are retained by this sieve and can be kept for later examination. The suspension is next passed through a 250 μm mesh sieve, which will retain larger spores and some hyphae, and finally through a 100 μm mesh sieve. The material held on the sieves should be vigorously washed with a spray of water. The majority of spores will be held on the 100 μm sieve and can be examined most easily by resuspending them in a small volume of water and filtering ths suspension through pieces of fine woven cloth such as 'Dicel' or 'Tricel', supported on a sieve. The yellow to brown spores deposited on the cloth can be seen and counted under a low power (× 20–×40) stereoscopic dissecting microscope.

Since species of V–A mycorrhizal fungi can be identified from their spore characters with some confidence and the hyphae of these fungi can also often be recognized, the method just described has given much information about the occurrence and distribution of different species. For example, a species of *Acaulospora* with honey-coloured spores has been found in soil in England, Scotland, New Zealand, Australia, Pakistan, South Africa and North and South America. Certain other species studied appear to be equally widespread, but some are more restricted in occurrence. It is likely that the more widespread species have been distributed throughout the world with planting material of cultivated plants. Geographically more restricted species may be those that have a narrower temperature range or possibly a more restricted host range.

Another fact that has emerged using the wet sieving and decanting technique is that hyphae of V–A mycorrhizal fungi are sometimes the major hyphal type in a soil. It is likely, therefore, that V–A fungi are often also amongst the most important, in terms of nutrient cycling in the soil. Although recent work has shown that these fungi may make limited saprophytic growth in soil, it is probably generally true that their hyphae will not remain alive for long unless connected to a live root, although spores can remain viable for as long as three years. Since V–A fungi are for practical purposes not culturable, studies of soil fungi based on isolation techniques give no evidence of their presence.

Several studies have shown that there are marked seasonal fluctuations in the numbers of spores that can be recovered from soil and the amount of root infection. In wheatfields at Rothamsted Experimental Station in southern England spore numbers remained low from December to June but then increased rapidly, and declined again from September onwards. The amount of root infection was small up to May but then increased to a maximum in September. This may indicate that V–A mycorrhizas were not an important factor in this particular soil in the growth of wheat plants during the crucial period of rapid Spring development. It would however be unwise to generalize from one set of observations such as these.

Evidence for the effects of manurial treatments on spore numbers and the amount of root infection is somewhat conflicting. In Rothamsted wheatfields all the plots without nitrogen added had more spores and more mycorrhizal root infection than plots with nitrogen added in the form of calcium nitrate. Numbers of spores increased when phosphate was added to the soil of plots with uninoculated maize plants. If the plants were already mycorrhizal when planted, adding phosphate had the opposite effect on spore numbers. These effects were probably partly due to the relative growth of the maize roots in plots with and without phosphate, but are not easy to interpret.

3.4 Fungi involved

Although there is still some uncertainty about the systematic position of fungi forming V–A mycorrhizas, most workers now accept that they are all members of the family Endogonaceae in the order Mucorales of the class Zygomycetes. However, the first observation of some of the fungi of this group were of their underground sporocarps. Because of some superficial resemblance to truffles, it was thought that these fungi were related to other members of the Tuberales of the class Discomycetes. The species of the Endogonaceae involved in mycorrhizal relationships differ from most of the Mucorales in not forming asexual spores in sporangia; while the sexually produced zygospores occur only in one genus. Gerdemann and Trappe (1974) recognized seven genera as belonging to the Endogonaceae and one more genus has now been described. These are listed below.

Endogone   This is the generic name that was once used for all the V–A fungi. As now defined it comprises species that form zygospores in sporocarps by the fusion of two gametangia. Three species are known to be mycorrhizal; they form sheathing mycorrhizal infections of certain tree species. Unlike the fungi belonging to all the other genera of the Endogonaceae, *Endogone* species have been isolated and grown in pure culture.

Gigaspora   Azygospores, similar in appearance to zygospores but not resulting from fusion of gametangia, are formed singly in soil on swollen cells at the end of hyphae. Thin-walled vesicles are formed on hyphae in soil. Mycorrhizas formed by species of *Gigaspora* have arbuscules and hyphal coils; vesicles if present are small.

Acaulospora   Azygospores are formed singly in soil on the sides of the stalks of thin-walled vesicles. Known species form V–A mycorrhizas.

Glomus   Chlamydospores are formed terminally, either singly, in clusters on hyphae, or in sporocarps. This genus includes the commonest V–A mycorrhiza formers and has 30 known species.

Sclerocystis   Rather similar to *Glomus* with chlamydospores formed in sporocarps in soil. The chlamydospores are arranged in a single layer and radiate from a central knot of hyphae. Species form V–A mycorrhizas.

Modicella   Species form terminal sporangia in sporocarps. Sporangial species of *Modicella* show some resemblance to a genus of common soil saprophytes, *Mortierella*.

Entrophspora   This is a recently described genus with one species (*E. infrequens*) and similarities with *Acaulospora*.

Glaziella   Chlamydospores are formed in large hollow sporocarps.

There is no evidence at present that any members of the last three genera are mycorrhizal.

4 Ericaceous Mycorrhizas

4.1 Introduction

The Ericales is a remarkably successful order of plants, occurring in almost all parts of the world. In both the Northern and the Southern hemispheres there are large areas which are dominated by ericaceous plants. Much of upland Britain is a good example of such areas, with relatively unproductive heathland often dominated by *Calluna vulgaris* and species of *Vaccinium* and *Erica*. It now seems likely that the success of such plants in these areas is due to specialized mycorrhizal associations common to most, if not all, ericaceous plants. Such associations may be of special importance in alpine regions where growing seasons are short. In typical ericaceous mycorrhizas the roots are penetrated intracellularly by septate fungal hyphae. This penetration is usually confined to the outer layer of cortical cells. Although there is no fungal sheath, a weft of hyphae is usually visible around the root. Infected rootlets appear externally to be non-mycorrhizal. Mycorrhizas of this type are typical of *Erica*, *Vaccinium*, *Rhododendron* and *Calluna*, all members of the Ericaceae, and are termed 'ericoid'.

Contrasting with these are the mycorrhizas of *Arbutus* and *Arctostaphylos*, also in the Ericaceae, *Monotropa* in the Monotropaceae and *Pyrola* in the Pyrolaceae. All these families belong to the order Ericales. The mycorrhizas of the four genera mentioned have a well-developed fungal sheath and a Hartig net from which extensive host cell penetration occurs. They are termed 'arbutoid' and resemble typical sheathing mycorrhizas of forest trees, perhaps forming a connecting link between the ecto- and endo-mycorrhizas.

The occurrence of such contrasting types of mycorrhizas within the Ericales is not really surprising when one considers the range of root system in this group. At one end of the scale the shrubs of the Ericaceae possess a system of very fine roots (known as 'hair roots') with typical endomycorrhizal infections. In contrast, the root systems of the closely related *Arbutus* and *Arctostaphylos* are differentiated into long and short roots similar to those of forest trees. This difference is reflected in their mycorrhizas as described above.

Pyrola and *Monotropa* have mycorrhizas somewhat resembling those of *Arbutus*. However, in *Monotropa*, representing the greatest contrast with the 'ericoid' type, there is a thick compact mycelial sheath and a Hartig net one cell deep in the cortex with much less intracellular invasion. Structurally these mycorrhizas closely resemble typical tree mycorrhizas but they function quite differently. To understand this it is necessary to consider the mode of nutrition of the Ericales. Plants in the Ericaceae are fully photosynthetic and have a mutualistic symbiotic relationship with their fungal partner. In the Pyrolaceae and Monotropaceae there is a full range from species of *Pyrola* that have green

Table 4–1 Characteristics of genera in the Ericales and their relationship to mycorrhizal type.

Family	Genus	Leaf character	Root form	Mycorrhizal type
Ericaceae	*Calluna*	Green, well developed	Very fine roots	Ericoid
Ericaceae	*Arbutus*	Green, well developed	Long and short roots	Arbutoid
Ericaceae	*Arcto- staphylos*	Green, well developed	Long and short roots	Arbutoid
Pyrolaceae	*Pyrola*	Green, well developed	Fibrous roots	Arbutoid
Pyrolaceae	*Pyrola*	Green but reduced, or lacking	Fibrous roots	Arbutoid
Monotrop- aceae	*Monotropa*	Achlorophyllous (yellow brown scales)	Clusters of fleshy roots	Arbutoid

leaves and are photosynthetic, through other *Pyrola* species whose leaves develop late and may be virtually non-functional, to *Monotropa* which is totally achlorophyllous. Those plants that are partially or totally lacking in photosynthetic ability are more or less dependent on their fungal partner. As with orchids (Chap. 5), the higher plant obtains its carbon and energy requirements from the mycorrhizal fungus on which it could be considered as parasitic. In *Monotropa* the relationship is further complicated by the fact that the mycorrhizal fungus forms a biotrophic relationship with a second higher plant, a tree. In this tripartite relationship *Monotropa* obtains at least the majority of its carbon and energy needs from the tree via the shared mycorrhizal fungus. Table 4–1 shows some characteristics of typical members of the Ericales and their relationship to mycorrhizal type.

4.2 Ericoid mycorrhizas

Structure   The distribution of the mycorrhizal endophyte in roots of *Calluna vulgaris* plants can be easily studied by carefully digging up some seedlings, preferably less than a year old, with as much of the root system and adjacent soil as possible. Adhering soil is gently washed off the roots in a stream of water, the whole root system stained (see § 3.2) and examined under a low power microscope.

The primary roots of *Calluna vulgaris* grow more or less vertically in the soil and give rise at intervals to secondary roots which terminate in extremely fine hair roots. The young hair roots have a very simple structure, consisting of a single layer of large cortical cells surrounding a very slender central stele. The roots lack both an epidermis and root hairs. It is within the hair roots that the heaviest infection occurs. The endomycorrhizal fungus grows loosely over the hair root surface and the septate hyphae penetrate the cortical cells, usually at several points in each cell, filling them with coiled hyphae. The apical region of the hair root remains free from infection and the stele is never penetrated. Transverse sections strikingly show the great quantity of fungus within the cortical cells (Fig. 4–1a). As much as 80% of the total root volume of heavily infected hair roots can be occupied by the endophyte. It is difficult to imagine any living tissue being unaffected when it is penetrated by such large amounts of fungus. The relationship between the intracellular hyphae and the host appears to be one of biotrophy rather than parasitism. Two important criteria of biotrophy are minimal damage to host tissue and disturbance of host nuclei. Invasion by the fungal endophyte results in the nuclei of infected cells becoming enlarged. Similar changes quickly follow plant infection by biotrophic parasites such as the rust fungi. After a period of active interaction between the hyphal coils and the host cytoplasm, degeneration of both partners occurs almost simultaneously; unlike the situation in V–A and most other mycorrhizas where fungal death precedes the death of the host cells.

Although surface hyphae do occur on older roots, infection of the cortex becomes progressively less from younger to older parts of the root system. This may be at least partly due to the older roots shedding their cortex and developing a corky bark. In natural conditions the intensity of infection varies between plants and changes seasonally. In the spring, active infection becomes evident. The intensity of infection increases to a maximum about mid-summer and then declines as the hair roots cease to grow. This basic pattern found in

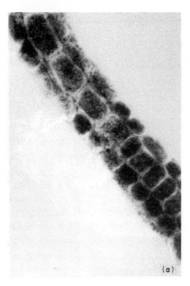
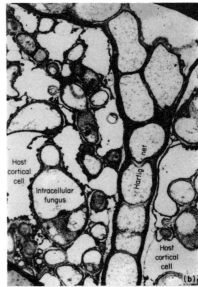

Fig. 4–1   (a) *Erica* root showing cortical cells containing mycorrhizal fungal hyphae (× 245); (b) transmission electron micrograph of transverse section of *Arctostaphylos* mycorrhiza with Hartig net and intracellular hyphae (× 3920). ((b) By courtesy of Dr J.A. Duddridge.)

Calluna also occurs in some other ericaceous genera. In *Vaccinium* and *Rhododendron* however, there may be as many as four layers of cortical cells around the stele of the stouter hair roots.

Ecology and fungi involved   Attempts to identify the fungal component have met, until recently, with little success. Early reports which implicated *Phoma* spp. have been largely discounted as infection experiments with numerous strains of *Phoma radicis* have failed to produce typical mycorrhizas. However, mycorrhizas similar to those that occur naturally have been formed when seedlings have been inoculated with the slowly-growing dark fungus isolated from *Calluna* roots and now described as a new species of the Discomycete genus *Pezizella* (see § 6.4). Although there are similarities between *Pezizella ericae* and isolates from other members of the Ericaceae and from various parts of the world, it seems rather unlikely that only one fungal species is involved, but more a group of very similar species. In addition, a somewhat tenuous link between *Clavaria*, a basidiomycete, and members of the Ericaceae, including azaleas and rhododendrons, has been recently established using the immunofluorescent staining technique.

4.3   Arbutoid mycorrhizas

4.3.1   *Arbutus and Arctostaphylos*

Structure   As mentioned earlier, the root systems of *Arbutus* and *Arctostaphylos* are differentiated into long and short roots similar to those of forest trees, and the short roots become invested by a sheath so that externally they look like true sheathing mycorrhizas. In addition, a Hartig net is formed, although it is usually restricted to the outer layer of cells. However, in sharp contrast to sheathing mycorrhizas of trees, in the arbutoid type the fungus penetrates the outer cortical cells filling them with hyphal coils. These features can be seen in Fig. 4–1b.

Ecology and fungi involved   Most fungal partners of arbutoid mycorrhizas, including those of *Arbutus* and *Arctostaphylos*, are probably basidiomycetes and very likely the same species that form sheathing mycorrhizas of forest trees.

4.3.2   *Pyrola and Monotropa*

Structure   *Pyrola* has a fibrous root system while *Monotropa* forms clusters of fleshy secondary roots. Their mycorrhizas rather resemble those of *Arbutus*. *Pyrola rotundifolia* is an evergreen creeping perennial that grows in bogs, fens and woods and has roots enveloped by a thick and compact fungal sheath. A typical Hartig net is formed, and intracellular hyphae develop within the cortical cells. Considerable penetration leads to the host nucleus becoming enlarged and lobed, and an increase in host cytoplasm. Some *Pyrola* species have either reduced or no leaves, but they have been little studied.

Monotropa hypopitys contrasts with species of *Pyrola* – it is an achlorophyllous herb that does not develop normally unless in direct contact with the hyphae radiating from the sheathing mycorrhizas of a range of woodland trees including beech, pine, spruce, fir, oak and cedar (Fig. 4–2a). It is often the only higher plant growing in the herb layer in deep forest shade. *Monotropa's* reliance on a tree for its supply of nutrients via a shared mycorrhizal fungus does not seem surprising on examination of its underground root system which is intimately interwoven with the roots of a tree (Fig. 4–2b). *Monotropa* roots are invested in a dense fungal sheath from which hyphae spread into the soil, either individually or as cords. Hyphae grow between the cells of the cortex to form a Hartig net similar to that of sheathing mycorrhizas but also penetrate the outermost cortical cells by means of fungal pegs which at first become enclosed by the host cell wall. As growth continues, extensive invaginations of the peg wall are formed, similar to those seen in specialized transfer cells (Pate and Gunning, 1972). Transfer cells are formed in a wide variety of plants when there is need for rapid and intensive short distance solute transport. The fungal pegs develop during the period of flower production. When the flowering shoot of *Monotropa* is fully developed and mature, the pegs then lyse and the cortical cells become invaded by fungal hyphae from the sheath. The senescence of the flowering scape follows.

Ecology and fungi involved   The fungal associates of the Pyrolaceae have not been determined with certainty to date. One fungus isolated from the roots of *Monotropa* was tentatively identified by comparison with pure cultures as a species of *Boletus*. Circumstantial evidence suggests that *Boletus subtomentosus* and *Boletus chrysenteron* are possible associates. The fact that *Boletus* spp. form mycorrhizas with a wide range of tree species may help to explain why *Monotropa* is associated with so many different tree species.

Fig. 4–2   (a) Plants of *Monotropa hypopitys* ($\times$ 0.4); (b) *Monotropa* and tree roots with surrounding hyphal cords ($\times$ 5.3). ((b) By courtesy of Dr J.A. Duddridge.)

5 Orchid Mycorrhizas

5.1 General description

The Orchidaceae is a unique family of plants that has excited the wonder and interest of scientists and non-scientists alike for centuries. Apart from their showy and bizarre flowers and often fascinating pollination mechanism, striking features of orchids include their minute and usually undifferentiated seeds, their tendency to saprophytism and the related fact that for part or all of their life cycle, almost all in natural conditions have an obligate dependence on infection by mycorrhizal fungi. The intimate involvement of fungi in normal orchid development was discovered and studied by the French botanist Noel Bernard during the first decade of this century. Orchid growers already knew that raising orchids from seed was a difficult and uncertain business. Bernard's experiments with aseptic seeds showed that these would not germinate, in contrast with seeds placed near parent plants, that could easily be colonized by any fungi associated with the mature plants. At first his attempts to isolate fungi from seedlings or orchid roots failed, but eventually he obtained cultures that stimulated germination of seeds of some orchid species. Although his researches showed Bernard that certain fungi were essential for the normal development of orchid seedlings, he regarded these fungi as parasites. He also developed a theory that the fungi were responsible for the formation of tuberous structures by orchids as well as some other plants.

Characteristically, the mycorrhizal fungi form intracellular coils or less regular hyphal aggregates within host tissue – these structures are known as pelotons (Fig. 5–1). The pelotons usually persist for a limited time, eventually collapsing and degenerating. However, degeneration of the pelotons may occur chiefly in deeper tissues, with those in more superficial tissues remaining apparently healthy and unlysed for a considerable time. There is no sheath or comparable structure in orchid mycorrhizas, but hyphae probably spread from the surface of infected organs through the surrounding soil.

5.2 The orchid life cycle

The relationship of orchids with their fungal symbionts is probably best understood by following the stages of development from ungerminated seeds to the mature plant. Depending upon species, each ripe seed capsule contains from a few thousand to several million seeds, each weighing no more than 0.3 to 14 μg, and consisting of a small group of similar cells enclosed by a thin seed coat or testa. The exact stage at which infection by the fungal symbiont becomes essential for further development varies with orchid species. In certain orchids, for example *Cypripedium*, germination is initiated by penetration by the fungus through the testa into cells at the suspensor end of

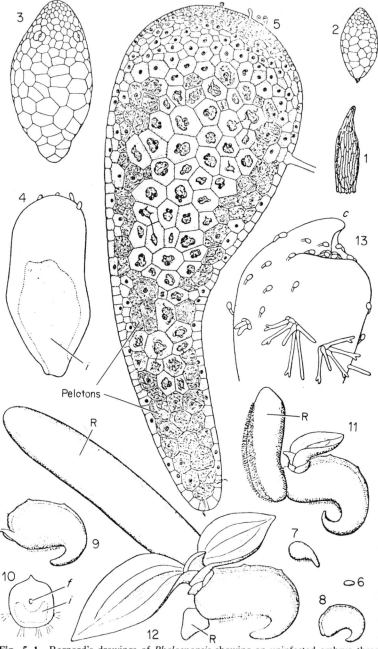

Fig. 5–1 Bernard's drawings of *Phalaenopsis* showing an uninfected embryo three months old (**3**) and another embryo fifteen days after infection by mycorrhizal fungus (**5**). Note the intracellular pelotons. (Reproduced from Bernard, 1909.)

the embryo. Fungal mycelium then spreads through the cells of the central part of the embryo. As tissue is colonized the embryo cells enlarge rapidly and the testa is ruptured. In other genera such as *Phalaenopsis* and *Cattleya*, germination starts in favourable conditions with absorption of water by the embryo, which swells and splits the testa. The very small quantities of nutrient reserves in the embryo are now used for limited growth. Further development cannot take place unless there is either direct absorption of organic compounds from the environment or infection by a suitable mycorrhizal fungus occurs. The fungus enters the embryo through its basal suspensor and rapidly penetrates most cells of the central part of the embryo, forming elaborate coils within these cells. Infection is usually accompanied by an immediate acceleration of embryo growth and by the beginning of organ differentiation (Fig. 5–1.)

At the other extreme from orchids like *Cypripedium* are some species of *Bletilla* that have rather well-differentiated seeds with significant reserves of carbohydrates. Such seeds will occasionally develop aseptically without obtaining any organic compounds from the environment, forming, in time, green leaves, and so becoming autotrophic. However, in the presence of a suitable fungus, growth and differentiation of the seedling is usually much quicker. By contrast, it is characteristic of most orchids that they may remain saprophytic for more or less prolonged periods. In extreme cases they never develop any chlorophyll but remain saprophytes for the whole of their existence. A British example of such a species is the Bird's Nest orchid, *Neottia nidus-avis* (Fig. 5–2), which grows in deep shade, especially in beech woods. Some wholly saprophytic orchids can reach a great size. *Galeola hydra*, although a terrestrial orchid, scrambles through forest trees for support and forms individual branches up to 2 cm in diameter and 16 m long. Amongst the majority of orchids that eventually become photosynthetic, there is much variation in the duration of the saprophytic phase, sometimes even within one

Fig. 5–2   Young plants of the Bird's Nest orchid, *Neottia nidus-avis* (× 0.4).

genus. *Orchis morio* may produce its first foliage leaf in the year following germination; in *O. ustulata* at least 10 years may elapse before the first leaf appears.

Let us now follow the changes that occur after infection of the germling or protocorm by its mycorrhizal fungus in orchids such as *Phalaenopsis*. Epidermal hairs start to develop and an apical meristem becomes active. Subsequently a central column of conducting tissue differentiates. It is in the protocorm that a regular sequence of fungal colonization followed by fungal degeneration is established. In the central part of the protocorm, following cell penetration, nuclear enlargement occurs and host cells appear physiologically active. Infection spreads from cell to cell so that the basal region of the protocorm becomes extensively infected. Growth of the intracellular hyphae leads to the formation of pelotons. These have a limited life and eventually collapse. The host cell, however, remains alive and active and may receive a new fungal penetration. Roots growing out from the protocorm may become independently infected from the substrate or by the fungus already in the protocorm. In these roots a similar pattern of colonization and degeneration occurs. As they develop, many orchids form modified storage roots or tubers in addition to absorbing roots. It is usual for the absorbing roots to be constantly re-infected from the substrate but for tubers to remain permanently free of fungal infection.

Bernard found that if small pieces of tissue were removed aseptically from orchid tubers and placed close to cultures of mycorrhizal fungi, colonization of the tuber tissue did not occur. On the contrary, the fungus was inhibited or killed by something diffusing from the tuber tissue. He also discovered that if the tuber tissue was killed by heating to 55°C or above it was no longer inhibitory. E. Gäumann and his colleagues working in Switzerland did similar experiments with tissue of *Orchis militaris*. They confirmed that killed tissue was not inhibitory and showed that the inhibitor was in fact not produced by the tissue until it was exposed to the fungus. Later they isolated and characterized the inhibitor which they named orchinol, a derivative of dihydroxy-phenanthrine. This compound is inhibitory to a range of mycorrhizal and non-mycorrhizal soil fungi. Further, its production is stimulated by all of the mycorrhizal fungi tested by Gäumann but only by certain of the non-mycorrhizal fungi. The significance of the discovery of orchinol is that it was the first example of a phytoalexin to be isolated and studied. It is now known that many, if not all, plants produce analogous inhibitors in response to challenge by potential pathogens. These inhibitors, or phytoalexins, have a wide spectrum of activity, but characteristically are less effective against fungi that are parasites of the plant producing them than against those fungi that are not.

Although in the majority of orchids mycorrhizal infection persists in the adult plant, which is therefore probably still at least partially saprophytic, this is not always the case. Some of the green terrestrial orchids, such as *Listera* and *Epipactis* spp, may be virtually fungus-free when adults, but they may have grown through an earlier prolonged totally saprophytic phase.

Bernard not only discovered that development in orchids normally depends on fungal infection, but he also showed that soluble organic compounds could at least partially replace fungi. For example, a decoction of orchid tubers known as salep stimulated seedlings. Sugars had a similar effect. Bernard's findings were confirmed and extended by L. Knudson in the 1920s. He succeeded in germinating and growing normal seedlings of a range of orchids using media containing sugars. One orchid, a *Laelia-Cattleya* hybrid, was grown right through to flowering without fungal infection. Despite his successes with many genera, Knudson totally failed to germinate seeds of some orchid species in sugar solutions. In several cases growth of seedlings in sugar media slowed down or completely ceased, but addition of various plant decoctions restored growth. It is now known that orchid seedlings may require certain vitamins, particularly nicotinic acid, amino acids and plant growth substances (e.g. auxins and kinins) for continued growth. Starch cannot replace sugar in media for asymbiotic orchid culture, but if media containing starch are inoculated with certain non-mycorrhizal fungi, a species of *Phytophthora* was used in one of Knudson's experiments, seedling growth may occur.

Studies of the asymbiotic growth of orchids are important for two reasons. Firstly, they have led to reliable methods, now commonly used by orchid growers, for raising orchids from seeds without the complications of inoculation with suitable fungi or dependence on chance natural infection. Secondly, asymbiotic culture experiments, by showing the nutrient requirements of achlorophyllous orchid seedlings, have been of great value in revealing the role of the mycorrhizal fungi. The fact that *Phytophthora* could stimulate growth of asymbiotic seedlings in a medium containing starch but not sugar suggests that fungal hydrolysis of starch resulted in a surplus of sugar that was directly absorbed by the orchid tissue. Under natural conditions other microorganisms would almost certainly use up any products of hydrolysis before significant amounts could be absorbed by the orchid. Normally growth stimulation does not occur until after penetration and colonization of orchid tissue by mycorrhizal fungi. This indicates that fungi supply carbohydrates by internal translocation through their hyphae rather than by external diffusion to the orchid. It is still not certain how carbohydrates and other compounds are transferred from the fungus mycelium to the orchid tissue. Recent work shows that part of the transfer is probably from live hyphae and part occurs necrotrophically as the intracellular pelotons degenerate.

5.3   The fungi involved

Since the mycorrhizal fungi of orchids are relatively easily isolated and grown in pure culture, it might be thought that their identification would also not be difficult. This had not been found to be so because, unless special methods are used, most remain sterile in culture. All that have been persuaded to produce perfect stages are Basidiomycetes and there is good reason to think that those that have remained sterile also belong to this group. A remarkable feature of orchid mycorrhizal fungi is that some at least are known to be destructive parasites of other plants. S. Kusano described an achlorophyllous orchid, *Gastrodia elata*, growing in Japanese oakwoods. When mature, this orchid consists of a massive system of underground rhizomes and tubers, giving rise to an inflorescence up to one metre high. Kusano noticed that, closely associated with the flowering tubers, there were always characteristic rhizomorphs of the notorious tree parasite *Armillaria mellea*. Other species of *Gastrodia*, growing in New Zealand, are almost completely underground and possess tubers covered with hyphae and rhizomorphs of *A. mellea* that penetrate adjacent tree roots on which the fungus is parasitic, e.g. *G. minor* has underground tubers that are found close to roots of the teatree (*Leptospermum scoparium*). The mycorrhizal fungus of this orchid forms, initially, a mycorrhiza-like infection of the teatree roots, with a recognizable sheath and even a Hartig net. The relationship is clearly parasitic as the root cortex disintegrates, while the orchid produces flowering shoots. There is at least circumstantial evidence that in these situations the orchid must be using nutrients obtained by the fungus from its tree host. The orchid could therefore be considered as a parasite of the tree by proxy, or an epiparasite.

Rhizoctonia solani (perfect stage *Thanatephorus cucumeris*) is another mycorrhizal fungus of orchids, strains of which are plant parasites with a wide host range. For example, isolates of *R. solani* obtained from hosts such as tomato and cauliflower, and shown to be pathogenic to them, formed a symbiosis with the British orchid species *Dactylorhiza purpurella*. It is now thought that more slowly-growing and nutritionally exacting rhizoctonias (e.g. *Tulasnella calospora*) may form a more stable symbiosis than faster growing species like *R. solani*.

It seems that the relationship between orchids and their mycorrhizal fungi may be rather precariously balanced. A change in environmental conditions can result in the previously symbiotic fungus becoming actively parasitic on its hosts. One series of experiments showed that the carbon sources being utilized by the mycorrhizal fungus can determine whether or not it becomes parasitic. With cellulose as the carbon source for the fungus, protocorms of a tropical orchid grew and remained in a healthy condition for 3 months. With glucose rather than cellulose as carbon source, many protocorms had become parasitized after 3 months, their growth had slowed and tissue browning had started.

As in the relationship of sheathing mycorrhizal fungi and their hosts, so with orchid mycorrhizal fungi there is a wide variation in the degree of specificity that occurs. Some fungi can form a successful symbiosis with a wide range of orchids, while others are much more specific. An example of a rather specific association is that between *Diuris* and the fungus *Tulasnella calospora*, studied by J. H. Warcup in Australia. All isolates from 28 plants were of *T. calospora*. In synthesis experiments using five species of *Tulasnella* and *Sebacina vermifera*, all fungi isolated from orchids, only *T. calospora* was fully effective in promoting the growth of *Diuris*. The association between *Dactylorhiza purpurella* and its endophytic fungi is much less specific. Experiments have shown that at least eight different fungi can infect this orchid and form a good symbiotic association. It seems that ability to infect orchid tissues and to produce intracellular hyphal coils is not necessarily an indication of an effective symbiosis. There is in orchids, as in sheathing mycorrhizas, probably a spectrum of effectiveness. In considering specificity it is important therefore to determine not only ability to infect but also the response of the orchid, which in turn may be affected by the nature of the medium.

6 Isolation, Culture and Nutrition of Mycorrhizal Fungi

6.1 Sheathing mycorrhizal fungi

6.1.1 Isolation from sheathing mycorrhizal roots

Although they are abundantly associated with most temperate trees, the distribution and relative frequency of sheathing mycorrhizal fungi, particularly those that do not fruit regularly or at all, are poorly known. This is partly due to the rather small number of studies that have been made, but also to the difficulties of isolating these fungi from roots and identifying them. Some species will not grow on any laboratory media so far tested, and must be regarded at present, like the V–A fungi, as obligate symbionts. Even when isolation has been achieved, there remain the problems of identification and proving, by synthesis of the mycorrhiza, that the fungus isolated is indeed mycorrhizal. Close and repeated association of fruit bodies with particular kinds of mycorrhizas, particularly if hyphal connections can be traced, will give strong presumptive evidence of identity. This will be strengthened if microscopic characters of the hyphae associated with the sheath and fruit body correspond (see § 2.1). For proof of identity, isolation both from fruit body and mycorrhiza and successful synthesis are necessary.

Most mycorrhizal workers have now adopted similar methods for isolation of mycorrhizal fungi from mycorrhizal short roots, although there are variations in detail. Pieces of long roots bearing short roots are carefully removed from the soil and immediately placed in plastic bags to prevent desiccation. Summer and autumn are probably the best seasons for isolation but successful isolations can be made from mycorrhizas collected at any time of the year.

The aim in attempting isolation is to remove as many contaminating organisms as possible with minimal damage to the mycorrhizal fungus. The following procedure is usually successful when isolation is possible. Pieces of long root about 10–20 mm long with short roots attached are freed of any large soil crumbs or pieces of organic matter and shaken for 10–15 min in small bottles of sterile water with a detergent (e.g. 0.003% (v/v) Tween 80) by hand or in a shaker, followed by a further 10–15 min shaking in one or two changes of sterile water. Various agents have been used as surface sterilants; 30% (v/v) hydrogen peroxide is one of the most effective. Washed roots are immersed in the hydrogen peroxide for 10–15 sec and then transferred to a few ml of sterile water in a Petri dish. Mycorrhizal roots 2–4 mm long are now cut from the long roots with sterilized scissors or scalpel and placed on the surface of a suitable isolation medium (p. 32). Antibiotics (e.g. streptomycin, penicillin or aureomycin) are often added to the isolation medium to suppress bacterial growth. The selective fungicide Benlate may be used if the mycorrhizal fungus to be isolated is a basidiomycete, but will suppress growth of ascomycetes and deuteromycetes, some of which form sheathing mycorrhizas.

The isolation plates are incubated at about 20°C and the roots on them examined daily through the bottom of the Petri dish, with a × 10 objective, for emerging hyphae. Usually fast-growing mycelia, particularly if they start sporing, are contaminants. However, species of *Rhizopogon*, *Suillus* and *Paxillus* grow out very quickly, often within a few days of plating. Most mycorrhizal fungi grow very slowly and may not even start to grow out of mycorrhizas for 10 or more days. Any fungi growing slowly and bearing clamps are very likely to be mycorrhizal. Others that grow slowly and often have narrow hyphae should be kept as they may possibly be mycorrhizal species.

6.1.2   Isolation of sheathing mycorrhizal fungi from sporocarps

Although some fungi, such as *Cenococcum geophilum*, do not produce sporocarps, the most successful and therefore most commonly used method of isolating mycorrhizal fungi, especially if named isolates are required, is from their often large and distinctive fruit bodies. Isolation from pileal or glebal fragments of a young fleshy sporocarp is successful in a higher percentage of species than by any other means. This is because there is far less risk of contaminating organisms getting into the sporocarp tissue and reducing the chance of successful isolation than when using mycorrhizal roots. To isolate from toadstools, the young sporocarp is first cut into cap and stipe and the latter discarded as it is often colonized by contaminants. The cap is then snapped in half, taking great care that nothing touches the newly exposed surfaces, to reveal the fleshy cap cortex or glebal tissue. A very small fragment of this tissue is then removed with a cool, sterile needle or knife and placed on the surface of an agar medium (see p. 32 for suitable formulations). Although more difficult to handle, truffles and earthballs are treated similarly. Initial mycelial growth from the tissue fragment may become visible by 48–72 hours and should be inspected regularly for any possible contamination. When hyphae thought likely to be of the mycorrhizal fungus (see § 6.1.1) have extended 1–2 mm over the agar, they can be transferred to fresh medium. The following are examples of mycorrhizal fungi that are relatively easy to isolate and culture – *Amanita*, *Boletus*, *Suillus*, *Leccinum*, *Paxillus*, *Laccaria*, *Pisolithus*, and *Scleroderma*.

6.1.3   Isolation from spores of sheathing mycorrhizal fungi

Airborne spores are assumed to serve as primary agents of dispersal and therefore inoculum for species of the mycorrhizal hymenomycetes and gasteromycetes. This is supported by recent studies in which spores of *Thelephora terrestris* and *Rhizopogon luteolus* have been successfully used for mycorrhizal synthesis. Although both these fungi grow well on nutrient media, their basidiospores, like those of most other basidiomycetous mycorrhizal fungi, germinate poorly, if at all, on laboratory media. This suggests that an activator for basidiospore germination, perhaps produced by living plant roots or other microorganisms, is involved in nature. Recent work has shown that a number of compounds produced by plant roots and by other soil organisms, including yeasts, may indeed stimulate spore germination, that inhibitors as well as stimulatory factors may be involved. Contaminant-free spores can be collected by attaching small pieces of hymenium-bearing surface (gills or pores) from fresh sporocarps to the lid of a sterile Petri dish with sterilized lanolin. The spores are allowed to drop for 10–16 hours onto an agar surface, after which the lid is replaced by another sterile lid and the plate incubated. As already suggested, this method will give only limited success.

6.1.4   Culture and nutrition of sheathing mycorrhizal fungi

Most sheathing mycorrhizal fungi that have been cultured grow best on media containing simple carbon sources such as glucose or mannose; more complex carbohydrates are less likely to be utilized. Certain species, however, do possess the necessary enzymes to utilize more complex carbohydrates including the sugar alcohols, mannitol and sorbitol, the disaccharides, cellobiose, maltose and sucrose as well as the polysaccharides, starch, pectin and cellulose. The presence of a small amount of starter glucose (0.1 g l^{-1}) in the medium greatly enhances hydrolysis of the more complex sugars by inducing the formation of 'adaptive' enzymes, including cellulases and amylases. In nature, it would appear that as the fungus obtains its carbon requirements from its host, these enzymes are suppressed. As most sheathing mycorrhizal fungi are not as efficient at breaking down carbon compounds as their litter-decomposing counterparts, they are poor competitors and probably capable of little or no saprophytic growth. The dominant nursery fungi, however, appear to be less fastidious in their carbon nutrition and thus perhaps better adapted to survive in the soil in the absence of a suitable host.

In contrast, the nitrogen (N) nutrition of sheathing mycorrhizal fungi appears to possess similarities with other soil fungi. They usually grow best on media containing ammonium salts (e.g. ammonium tartrate) but some benefit from one or more amino acids. Most higher plants can use both ammonium N and nitrate N although there may be preferences for one or the other. Some mycorrhizal fungi (e.g. *Amanita muscaria*) appear not be be able to utilize nitrate, although several, such as *Lactarius pubescens*, have now been shown to produce nitrate reductase and to grow on media in which nitrate is the sole nitrogen source. Thus the generalization that mycorrhizas cannot absorb nitrate efficiently appears unsound. It may be advantageous to select mycorrhizal fungi for tree inoculation on the basis of their ability to use both nitrate and ammonium ions.

Sheathing mycorrhizal fungi can also use a wide range of organic nitrogen compounds including many amino acids (e.g. glutamic acid) and more complex substances such as peptones and nucleic acids. Other sources of nitrogen, including casein hydrolyzate, malt extract and yeast extract, also stimulate vegetative growth. The ability of mycorrhizal fungi to release nitrogen from organic sources has led to the idea that these fungi could 'short circuit' the decomposition-mineralization cycle for plant nutrients.

Fungi also require a range of minerals; some, such as potassium and phosphorus, in larger quantity and others, such as manganese and zinc, in traces. In addition to these essential nutrients, many sheathing mycorrhizal fungi have been shown to be stimulated by, or dependent upon, certain B vitamins; thiamin and biotin being the major ones. For example the majority of *Boletus* species have been found to be heterotrophic for thiamin. Both thiamin and biotin should be included in synthetic culture media. A few fungi have been reported to be heterotrophic for inositol, nicotinic acid and pantothenic acid. Complexes of vitamins may either stimulate or depress growth.

The optimum temperature for most sheathing mycorrhizal fungi in culture lies between 18° and 27°C. The growth of many fungi ceases above 35° and below 5°. There are, however, exceptions which may be explained by the origin of particular isolates. For example, a strain of *Paxillus involutus* isolated from a valley grew at a minimum of 2° to 8°, whereas the minimum of a strain isolated from a mountain range was −2° to 4°. In contrast, *Pisolithus tinctorius* has an optimum growth temperature of 28° with an upper limit of 40°. The tolerance of relatively high temperatures by this fungus may explain why it is one of the primary symbionts on young seedlings growing on coal wastes in America where soil temperatures are known to reach 35°C to 65°C.

In general, sheathing mycorrhizal fungi are acidophilic; most species growing best in pure culture at pH 4–6, although certain species grow well at pHs outside this range (e.g. isolates of *Paxillus involutus* between pH 2.7 and 6.4). Many other mycorrhizal fungi may grow over just as wide a range, but it is probably wise to use media with an initial pH near 5.0 to suit as many fungi as possible.

As far as we know most sheathing mycorrhizal fungi are obligate aerobes and should therefore be grown in the presence of sufficient oxygen. Certain fungi are extremely sensitive to water stress, although *Cenococcum geophilum* has been found to be particularly tolerant to artificial drought conditions and as a result has been recommended for inoculating trees to be planted in dry areas.

In an attempt to provide the chemical and physical requirements favouring as wide a range of sheathing mycorrhizal fungi as possible, two culture media have been developed for their isolation and growth. These are Hagem's and Melin-Norkran's media and various modifications of them (Table 6–1). Many

Table 6–1   Media for sheathing mycorrhizal fungi.

Modified Melin-Norkran's agar			*Hagem's agar*		
KH_2PO_4	0.5	g	KH_2PO_4	0.5	g
$(NH_4)_2HPO_4$	0.25	g	NH_4Cl	0.5	g
$MgSO_4.7H_2O$	0.15	g	$MgSO_4.7H_2O$	0.5	g
$FeCl_3$ (1.0% sol.)	0.5	ml	$FeCl_3$ (1.0% sol.)	0.5	ml
$CaCl_2$	0.05	g	Glucose	5.0	g
NaCl	0.025	g	Malt extract	5.0	g
Glucose	10.0	g	Thiamine HCl	50	μg
Thiamine HCl	100	μg	Agar	10	g
Agar	10	g			
Distilled water to	1000	ml	Distilled water to	1000	ml

other media have also been tested with some success. Since one medium will quite often give better results than another, it is a good idea in initial isolation attempts to try at least two different media. Isolation of mycorrhizal fungi is sometimes easier if diluted media, containing one third to one sixth of normal concentrations of nutrients, are used. This may help by reducing the rapid growth of any contaminating organisms. Generally solid media are better than liquid media especially for isolation. *Elaphomyces granulatus* however does not grow on agar media, but will grow very slowly in suitable liquid media. Clearly media for isolating and growing mycorrhizal fungi need much improvement as many fungi have still not been isolated, while others grow only very slowly even when they are isolated. Further information on synthetic media and culturing fungi is given by Deverall (1981).

6.2 V–A mycorrhizal fungi

The fungi forming V–A mycorrhizas represent an extreme form of adaptation to a symbiotic mode of life; they have not been successfully isolated and grown in pure (axenic) culture. The nearest approach has been to establish pure two-membered (monoxenic) cultures consisting of fungus plus a host seedling or root organ culture, growing together in symbiotic association on a defined medium. Mosse and Phillips (1971) have produced such cultures in the following way. Chlamydospores of *Glomus mosseae* were dissected out of sporocarps and surface sterilized in a solution of Chloramine T plus 200 mg l^{-1} streptomycin and a trace of detergent. The spores were transferred by pipettes to watch glasses containing the sterilizing solution and then rinsed in three changes of sterile water. Seeds of the clover species *Trifolium parviflorum* were surface-sterilized in H_2SO_4, rinsed thoroughly in water and sown on water agar in Petri dishes. On germination, seeds were transferred singly to slopes of agar in test tubes and allowed to grow in the greenhouse. When the seedlings had lateral roots, 10–15 *Glomus* spores were placed on the agar surface near the roots of each seedling. A typical medium that gave successful infections had the following composition in g l^{-1}: agar, 15.0; KNO_3, 0.5; $CaHPO_4$. $2H_2O$, 0.55; $MgSO_4$. $7H_2O$, 0.2; $CaCl_2$, 0.2; Fe EDTA, 0.007 Fe. When attempts were first made to obtain pure two-membered cultures of V–A fungi and host seedlings grown in Jensen's medium, it was found that, if contaminating bacteria were eliminated, infection was poor in this medium which contains much phosphate. Subsequent work showed that the presence of certain soil bacteria increased infection and that pectinase and EDTA also had this effect. In later work good infection resulted in media without bacteria, pectinase or EDTA, if less than 1.5 g l^{-1} monocalcium phosphate was included in the medium. It is not known whether bacteria play a significant role in the infection process in soil.

One consequence of the obligate nature of the V–A mycorrhizal symbiosis for the fungus is that it is impossible to study the physiology of the fungus independently of the host plant. In experiments in which responses to nutrients added to two-membered culture systems are observed it is difficult to distinguish between direct effects on the fungus resulting from fungal uptake from the medium and plant mediated effects. One promising, but experimentally difficult approach, is to isolate physically the substrate containing most of the external mycelium from the substrate containing the plant roots, making certain that the external mycelium is still connected to the roots. This is a similar technique to that used in some orchid mycorrhizal research. The method has been used to follow the movement of phosphorus (P) from the external environment via hyphae to the roots (see Chap. 8).

Experiments in which mycorrhizal roots have been exposed to $^{14}CO_2$ have shown that labelled photosynthate moves from the roots into the external mycelium. It has also been noticed that starch tends to disappear from infected root cells. However, the extent of movement of photosynthate from plant to mycorrhizal fungus is much less in V–A than in sheathing mycorrhizas. The photosynthate transferred to the external mycelium is incorporated in proteins and cell wall material and some is stored as organic acids and glycogen.

Rather little is known about the availability to V–A fungi of carbon and energy sources in the soil. It is likely that in an active symbiosis the fungus obtains most of its requirements for carbon from the host. However, experiments with ^{14}C have shown that mycorrhizal roots of onion have an increased ability to incorporate carbon from external sources such as sucrose, acetate and glycerol as well as from photosynthate. It is not certain whether these carbon compounds taken up externally are absorbed directly by the roots or via the external mycelium. There is also evidence that inositol and phytate may serve as both phosphorus and carbon and energy sources for V–A fungi.

Isotopic P has been used to follow the movement of P from the soil through the external mycelium to the host root. While in sheathing mycorrhizas P accumulates in the sheath, V–A mycorrhizas, possessing no sheath, cannot accumulate P to the same extent, but pass it directly to the host.

6.3   Orchid mycorrhizal fungi

Orchid mycorrhizal fungi are not usually difficult to isolate (see Chap. 4). Most grow quite rapidly on simple media, special precautions to restrict the growth of contaminants may not be necessary. However, the incorporation of a wide-spectrum antibacterial antibiotic such as aureomycin in the isolation medium does help to ensure that fungal isolates are free from bacteria. Successful methods for isolating orchid mycorrhizal fungi include plating out fragments of previously thoroughly washed and surface sterilized roots or proto-corms, and plating out groups of infected cells dissected from orchid tissue. Techniques similar to that described in § 6.4 for isolating the mycorrhizal fungus from *Calluna* are also used for isolating orchid mycorrhizal fungi.

Although most orchid mycorrhizal fungi grow well in culture they may not always be easy to distinguish from saprophytic fungi. One reason for this is that several common imperfect basidiomycetes do not have clamps and cannot therefore be immediately recognized as basidiomycetes. However, their hyphae do tend to have other characteristics that can be recognized with practice.

Mycorrhizal fungi of orchids are all nutritionally active, being able to use a much wider range of carbon substrates than sheathing mycorrhizal fungi. All that have been tested can use simple sugars, starch and a range of other carbohydrates. Probably of more importance in their symbiotic association with orchids is their ability to use complex compounds, particularly cellulose and, in some cases, lignin. This ability can result in a virtual short-circuiting of the normal carbon cycle. The carbon present in structural compounds of dead plant remains in the litter can pass as soluble carbohydrates through mycorrhizal fungi directly to orchids. As we saw in Chapter 4, there is some evidence that nutrients may flow from a diseased but living plant to an orchid through hyphae of a fungus that is simultaneously pathogenic and mycorrhizal. *Armillaria mellea*, the fungus reported to be involved in these relationships, is by no means always parasitic, it may live for long periods as a saprophyte on dead woody tissues. Strains of this fungus able to degrade cellulose and lignin produce a characteristic white rot. However, some mycorrhizal isolates of *Armillaria* degrade cellulose but not lignin and produce brown roots.

Most orchid mycorrhizal fungi that have been tested make very limited growth with ammonium or nitrate as nitrogen sources, but grow much better with amino acids such as asparagine and glycine or with urea. This may sometimes be due to stimulatory impurities that are associated with these compounds rather than to a direct effect of the nitrogen compounds. There is experimental evidence that several orchid mycorrhizal fungi grow better when provided with yeast extract or very small amounts of growth factors such as thiamine or p-aminobenzoic acid. It seems likely that under natural conditions the orchid provides its mycorrhizal fungus with growth factors, just as the fungus may supply the orchid with other growth factors that it needs.

We have hypothesized that soluble organic compounds (carbohydrates), released from insoluble plant residues by fungal enzymes, are transported through the mycelium to the plant host. What evidence have we that such transport through hyphae actually occurs? The most informative experimental studies of translocation in fungi have used a split culture plate method in which one part of a culture plate is separated from the other by a barrier that prevents diffusion. If an isotopically-labelled nutrient is added to one side of the plate and the isotope can subsequently be detected in hyphae on the other side of the barrier, translocation through the hyphae can be assumed to have occurred. Using this method, both *Rhizoctonia solani* and *R. repens*, known mycorrhiza formers, have been shown to translocate ^{14}C and ^{32}P absorbed from the substrate in labelled glucose and labelled potassium phosphate respectively. It is interesting that in some of these experiments the majority of the ^{14}C detected in mycorrhizal fungal hyphae has been in the compound trehalose, a common fungal disaccharide. A modification of the split culture plate method has also been used to show that both C and P can be translocated via *Rhizoctonia* hyphae through a diffusion barrier in to orchid seedling tissue.

The evidence now available suggests that orchid mycorrhizal fungi range from those that are quite good saprophytic competitors, and sometimes necrotrophic parasites, to more specialized and nutritionally exacting species. Some of the latter, like many sheathing mycorrhizal fungi, may be ecologically-obligate biotrophs, although probably most are better regarded as ecologically-facultative biotrophs.

6.4   Ericaceous mycorrhizal fungi

There have been many attempts to isolate and identify the mycorrhizal fungi from *Calluna* and other ericaceous plants. Some have been unsuccessful while others have yielded isolates with similar cultural characteristics, able to form typical mycorrhizas in synthesis experiments.

Of the different methods used to isolate the mycorrhizal fungus of ericaceous plants, those using tissue maceration seem to be most successful. Young infected hair roots collected during the Spring from an ericaceous plant such as *Calluna vulgaris* are placed in a muslin bag under running water for 2–4 hours to remove loose debris. To get rid of the remaining contaminating microorganism the roots are then carefully washed in about 20 changes of sterile distilled water. They are then placed in a sterile glass vial with sterile water and macerated with a glass rod to give a suspension of cells and groups of cells. A drop of this suspension is placed in a sterile Petri dish and molten cooled distilled water agar (0.5%) is added. The root cells are then dispersed in the agar by gently rocking the dish.

After incubation for about 3 days at 20°C single infected cells with hyphae radiating from them can usually be seen under the microscope. In many cases these hyphae can be seen to be growing from within a root cortical cell, confirming the endophytic nature of the fungus. Small blocks of agar containing cells with emerging hyphae are transferred aseptically to dishes containing 2.0% malt extract agar. The risk of contamination is considerably reduced by this method in which individual infected cells are dispersed in a non-nutrient medium. This is important as the endophyte is slowly growing.

After 4–6 weeks, cultures of the endophyte are grey to fawn and velvety. One strain of the fungus isolated from *Calluna vulgaris* roots is known to fruit when cultures are illuminated for several weeks, forming a ring of bright yellow apothecia around the culture margin. This made it possible to identify the fungus as *Pezizella ericae*.

The fungus grows well on a range of other common culture media besides malt agar and is nutritionally rather similar to sheathing mycorrhizal fungi. It can use simple monosaccharides and some complex polysaccharides (e.g. starch, pectin), but possesses only a limited capacity to degrade cellulose. The nitrogen nutrition of *Pezizella* is particularly interesting. While it can use the inorganic ammonium and nitrate ions as a source of nitrogen, it grows better when provided with certain organic nitrogenous compounds such as DL-alanine and glutamic acid. It can also use organic phosphate compounds as a source of phosphorus. These observations suggest that the mycobiont's nutrition may be of great advantage to its host plant, which usually grows in soil in which most of the nitrogen and phosphorus are present in organic forms.

Since the fungi forming arbutoid mycorrhizas are species that also form sheathing mycorrhizas (§ 4.3.1) their nutrition might by expected to be similar to that of most sheathing mycorrhizal fungi. That there is nothing special about the strains of fungi that form arbutoid mycorrhizas is confirmed by synthesis experiments; *Corticium bicolor* isolated from a Douglas Fir mycorrhiza, and *Cenococcum geophilum* isolated from a slash pine, can both form good mycorrhizas with Pacific Madrone (*Arbutus menziesii*).

7 Synthesis of Mycorrhizas

7.1 Sheathing mycorrhizas

Very much more is known about the environmental factors controlling the formation of sheathing mycorrhizas than any other type of mycorrhiza. These factors, therefore, are now discussed in some detail before a description of the synthesis of sheathing mycorrhizas in laboratory conditions is given.

7.1.1 Factors affecting sheathing mycorrhizal development under natural conditions

The process which leads to the successful association between host and fungal symbiont has already been described (p. 7). However the factors controlling this process are complex and involve a close interaction of higher plant, symbiotic fungus and environmental conditions. It is therefore not surprising that many theories concerning mycorrhizal formation have arisen, with most theories regarding the higher plant and its internal state as of paramount importance.

Carbohydrates The main factors influencing the susceptibility of the host's roots to mycorrhizal infections appear to be photosynthetic activity and soil fertility. In general, high light intensity or a relatively long photoperiod coupled with moderate soil fertility favour mycorrhizal formation. In contrast, low light intensity (below 20% of full sunlight) and very high soil fertility may result in few or no mycorrhizas being formed.

Early observations that the abundance of mycorrhizas on plant roots might be controlled by the availability of mineral salts led to the theory that mycorrhizal formation depended upon low nutrient concentrations within the plant's roots and high concentrations within the fungus; conditions typically found in infertile soils. Later it was suggested that nitrogen (N) and phosphorus (P) levels in the soil influenced mycorrhizal development indirectly by affecting the content of soluble reducing sugars in the roots. It was postulated that a high level of soluble carbohydrates in roots is a pre-requisite for symbiotic association. High concentrations of N and P promote root growth, and hence protein synthesis, in the plant, thereby decreasing the amount of available carbohydrates in the roots, and as a result reducing mycorrhiza formation. According to the carbohydrate hypothesis of E. Björkman light works in exactly the same manner, affecting mycorrhiza formation by its influence on the carbohydrate levels of roots. Björkman's hypothesis has been criticized, partly because of the analytical methods he used. Recent work in America, however, has shown that high levels of N and P decrease the soluble sugar concentration in short roots of loblolly pine and reduce their susceptibility to

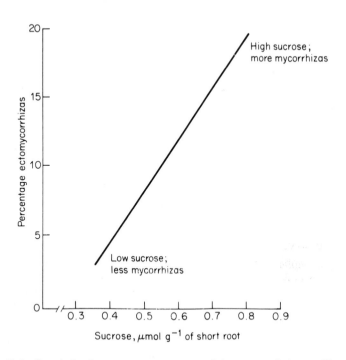

Fig. 7–1 Correlation between sucrose content of short roots of pine seedlings and percentage ectomycorrhizas formed after inoculation with *Pisolithus tinctorius*. (Redrawn from Marx, *et al.*, 1977.)

infection by *Pisolithus tinctorius* (Fig. 7–1), thus lending support to Björkman's hypothesis. However, the major sugar controlling infection was found to be sucrose, not a reducing sugar as Björkman had believed.

Auxins In addition to Björkman's hypothesis, several other theories concerning the factors which control mycorrhizal formation have been proposed. It is suggested that the conditions which permit and promote symbiotic association are controlled by a complex of both internal and external factors. Auxins probably have a central role in the formation and functioning of sheathing mycorrhizas (as mentioned briefly in Chap. 2). Anatomical features frequently shown by sheathing mycorrhizas are a swollen appearance, lack of root hairs and radial elongation of the outer cortical cells, the latter occurring in some species only. Similar changes in non-mycorrhizal roots of pine seedlings can be induced by culture filtrates of mycorrhizal fungi or by synthetic auxins. This suggests that sheathing mycorrhizal fungi produce extracellular auxins and that these auxins induce the characteristic mycorrhizal form. Several studies have shown that sheathing mycorrhizal fungi produce indole acetic acid (IAA) and other indole compounds when grown in axenic culture, thus lending support to this idea. It has also been observed that the mycorrhiza-like structures, induced either by fungal exudate or synthetic auxins are not permanent, but last only as long as the exudates or auxins are applied. However, the actual transfer of indole compounds or other hormones (e.g. cytokinin) from the fungus to roots has not yet been demonstrated and the amounts of auxin obtained in pure cultures may be too small to induce the characteristic morphogenic changes. This has led to some doubts about the fungus as the main source of growth substances and the alternative theory that it is plant tissue which produces a higher than usual concentration of auxin with the consequent morphological changes. It should, however, be borne in mind that the amounts of extracellular auxins produced in pure culture may be quite different from the amounts that the fungus produces when in association with the host plant.

As already mentioned, the inhibitory effect of high levels of N on mycorrhiza formation may be due to an increase in protein synthesis being followed by a decrease in sugar levels in the roots. Alternatively, this effect may be auxin mediated, as increasing N levels in cultures of mycorrhizal fungi can decrease the production of extracellular auxins. Another suggestion is that when N levels are high, auxins formed by sheathing mycorrhizal fungi may be converted into inactive compounds or might be destroyed by the host plant auxin oxidases.

Low light intensity has also been shown to inhibit mycorrhiza formation. Despite indications that the carbohydrate status of the root is of great importance, it seems that the influence of light may be more complex, involving additional processes. For example, even in the absence of mycorrhizas, reduced light intensity may greatly retard root growth, virtually stopping it at intensities below 20% of daylight. It has also been observed that with low light intensities IAA does not induce mycorrhiza-like structures in roots of aseptically-grown intact *Pinus strobus* seedlings, whereas it does at higher light levels. It follows that in plants grown at low light intensity in the presence of the appropriate mycorrhizal fungus, mycorrhizas might not be formed because of reduced root growth and the lack of response to fungal auxins.

Auxin stimulation and hence mycorrhiza formation also appears to be affected by the rhizosphere soil organisms. Many such organisms are known to accelerate auxin synthesis by supplying symbiotic fungi with extracellular substances, including vitamins (e.g. thiamin), while others produce auxins directly. The most active of these are species of *Pseudomonas* and *Azotobacter*. Additional substances which stimulate fungal activity are also present in plant root exudates. Several of these have already been mentioned, for example sugars, vitamins and hormones. However, much more information is needed concerning the full range of metabolites produced by host and fungus. For example, the roots of *Pinus sylvestris* exude a stimulatory metabolite termed the M factor which has not yet been identified but which can be replaced by diphosphopyridine nucleotide. By contrast, inhibitory substances may be produced by the host plant and accompanying rhizosphere organisms which reduce or completely arrest fungal growth. Examples of fungal inhibitors produced by rhizosphere microorganisms are gliotoxin and griseofulvin.

Temperature   Temperature profoundly influences the colonization of roots by mycorrhizal fungi. Although mycorrhiza formation may occur at relatively low temperatures, higher temperatures generally accelerate this process. At 16–18° the numbers of mycorrhizas on oak seedlings was nearly twice that on seedlings grown at 12–14°. However, wide ecotypic variation both within and between fungal species means that there are fungi adapted to a wide range of soil temperatures. *Cenococcum geophilum* is an example of a fungus that occurs over a wide range of climatic zones. *Pisolithus tinctorius* is favoured by moderate to rather high soil temperatures and can survive even higher temperatures (see § 6.1.4).

Soil temperature also affects the amount and composition of root exudates as well as the elongation and maturation of roots. Any change in root physiology could obviously affect colonization by sheathing mycorrhizal fungi.

Moisture and aeration   Most sheathing mycorrhizal fungi cease growth or grow very slowly when soil water is deficient. After dry periods of at least three weeks, the sheath and outer layers of mycorrhizas shrink and root growth stops. Once the soil has been remoistened, mycorrhizas regenerate from the renewed elongations of roots. *Cenococcum geophilum*, however, forms mycorrhizas and presumably continues to absorb nutrients under more severe moisture stress than most other fungi. Its characteristic black mycorrhizas are common in dry soils.

Excessive soil moisture can also severely affect mycorrhiza formation. In Finland's water-logged peat bogs, tree mycorrhizas occur only near the soil surface and are often dark and slender. In soils with excess water, poor aeration limits development of both the fungus and tree roots. The roots of some tree species seem better able to grow in conditions of partial anaerobiosis. For example, air spaces in the stele of roots of *Pinus contorta* appear to facilitate the movement of oxygen from the shoot to the roots and consequently maintain normal physiological activity even in very wet soils. Mycorrhizal fungi are strongly aerobic and will colonize tree roots only if sufficient oxygen is present.

pH   Most sheathing mycorrhizal fungi need an acid soil for good development and hence are inhibited in neutral to alkaline soils. This is probably primarily due to the direct effect of pH on fungal growth but indirect effects may also be important. For example, the inhibitory effect of alkaline soils on mycorrhizal formation has been ascribed to an increased availability of nitrogen resulting from enhanced nitrification rather than to the high pH itself. However, trees growing in alkaline soils may form abundant mycorrhizas. Recent work in France has shown that certain pine species may actually owe their lime tolerance to a symbiotic association with the appropriate fungus. Just as temperature optima of strains of a mycorrhizal fungus may differ so may their pH optima. Nine strains of *Paxillus involutus* grown in identical nutrients were found to have pH optima ranging from 3.1 to 6.4.

7.1.2   Synthesis of sheathing mycorrhizas

As shown in § 7.1.1, both physiological and environmental factors influence the success of the association between host and fungal symbiont. Knowledge of these influences has helped the development of techniques for synthesizing mycorrhizas in pure two-membered cultures, i.e. fungus plus plant but no other organisms. Synthesis of mycorrhizas in this way is necessary for fundamental studies including: (a) identification and confirmation of specific host/fungus associations; (b) the study of mycorrhizal physiology; and (c) comparison of the effects of different mycorrhizal fungi on a single or several host species.

Melin's original technique for achieving aseptic culture was to grow contaminant-free seedlings in flasks containing sand inoculated with a culture of a mycorrhizal fungus. This method has been developed in a variety of ways. Most workers now use mixtures of vermiculite and peat instead of sand. This provides better aeration and water holding capacity than sand and has been used successfully for the synthesis of sheathing mycorrhizas between many different fungi and tree species. A mixture of vermiculite and peat in the ratio of about 9:1 usually has a pH of 4.8–5.2, ideal for the formation of most mycorrhizas. Such a mixture is also excellent for growing many sheathing mycorrhizal fungi when a nutrient solution, often a modification of Melin-Norkran's solution, is added. Flasks or glass preserving jars are often used to contain cultures on vermiculite/peat for use in inoculation experiments. If, as is often the case, the experiments are to study growth responses to inoculation, surplus unused nutrients are leached out of the inoculum before it is used by rinsing in running water. Synthesis experiments have been done successfully in a variety of containers, ranging from quite small tubes to large flower pots. Choice depends on the tree species and the length of time for which it is to be grown. Formation of mycorrhizas, and even significant growth responses, often occur in less than 8 weeks. One disadvantage of vermiculite/peat cultures is that, because the medium is opaque, it is usually necessary to sample destructively to find out whether or not mycorrhizas have formed.

It is possible, at least with some tree species, to use an agar medium in glass or plastic tubes for synthesis experiments (Mason, 1980). This allows continuous observation of the roots throughout the experiments. There have been many studies on the effects of the culture medium on mycorrhiza formation. Unsuitable media may result in abnormal mycorrhizas, or even completely inhibit mycorrhiza formation. With a suitable, balanced nutrient medium mycorrhizas may develop quite quickly, for example 4–6 weeks with birch seedlings and only two weeks with spruce seedlings. Studies using the kind of methods just described have shown that, in addition to the physical and chemical environment, inherent genetically-controlled factors in the plant and fungus also determine the interaction that takes place between them. This is a situation analogous to that studied extensively in the legume nodule/*Rhizobium* symbiosis.

7.2   Vesicular-arbuscular mycorrhizas

Some aspects of the synthesis of V–A mycorrhizas have already been discussed in § 6.2. As we discovered, V–A fungi can be regarded as obligate symbionts. A consequence of this fact is that cultures of the fungi can be achieved only by synthesizing the symbiosis. A procedure for synthesizing V–A mycorrhizas in test tubes has been described.

The first successful experimental plant inoculations were made by B. Mosse, who discovered sporocarps of a V–A fungus in hyphal connection with the roots of mycorrhizal strawberry plants. She used spores dissected from sporocarps as inoculum to establish two-membered cultures. Sporocarps were either not treated or were surface-sterilized with mercuric chloride or Chloramine T. These treatments reduced rather than eliminated contamination of sporocarps. Spores were surface-sterilized with the same agents, and then thoroughly washed in sterile distilled water. Sporocarps and individual spores produced infections of strawberry plants growing in partially-sterilized and autoclaved composts. In many cases more sporocarps formed around the roots of infected plants. Experiments like these, using soil or compost, eventually led to the development of the pure two-membered culture methods described in § 6.2.

Since axenic culture of V–A fungi is not at present feasible, inoculum for use in experimental work is usually obtained from previously inoculated roots and the surrounding soil. This should ensure that the same species of V–A is used in a series of experiments.

7.3   Orchid mycorrhizas

Experimental studies have shown that synthesis of orchid mycorrhizas by inoculating aseptic seedlings with appropriate cultures is not usually difficult. The following procedure, with various modifications, has been found successful for several orchid and fungus combinations. Orchid seeds are surface-sterilized by shaking for 20 minutes in hypochlorite solution (5.0% w/v 'Domestos' works well), and then rinsed in several changes of sterile water. The seeds are then spread aseptically over the surface of agar slopes; up to 200 or more being sown on each slope. A medium that usually gives good results is modified Pfeffer solution agar containing (w/v) 0.08% $Ca(NO_3)_2$; 0.02% KH_2PO_4; 0.02% $MgSO_4$. $7H_2O$; 0.02% KNO_3; 0.01% KCl and Fe, Zn, Cu and Mo in traces, supplemented with 0.05% $(NH_4)_2SO_4$ for certain species and solidified with 1.0% agar. Ball-milled cellulose powder (1.0% w/v) is an excellent carbon source for most orchid mycorrhizal fungi tested. It results in more rapid development of protocorms than when a sugar such as glucose is used.

The mycorrhizal fungal inoculum is added as a disc cut from an agar culture. Infection and formation of pelotons can occur quite rapidly. The process has been observed microscopically and the various stages timed with protocorms of

Dactylorhiza purpurella and species of the fungal genus *Ceratobasidium* in slide cultures. Epidermal hairs of protocorms are penetrated in approximately 15 hours and peloton formation occurs in another 15 hours. Collapse of pelotons starts about 10 hours after their formation. Stimulation of protocorm growth follows infection quite rapidly and may start even before the collapse of pelotons.

7.4   Synthesis of ericaceous mycorrhizas

7.4.1   Ericoid mycorrhizas

As mentioned in § 6.4, the common ericoid mycorrhizal fungus can be quite easily isolated and cultured, and its identity is now known. Cultures of this fungus, *Pezizella ericae*, can be used to inoculate seedlings, making possible detailed studies of the physiology and ecology of the mycorrhizal association. Most successful synthesis procedures have made use of sand or agar as a culture medium. Inoculum of *Pezizella* is grown in agar, or probably better, in a liquid nutrient medium. Before use it is washed free of surplus nutrients and macerated to form a mycelial slurry. Seeds of *Calluna*, *Vaccinium* or *Erica* spp. are surface-sterilized and then germinated aseptically on water agar. The seedlings are transferred to culture vessels when their cotyledons have fully emerged. In the agar culture method the seedlings are planted in a thin layer of sterilized mineral soil overlying a bed of water agar in a Petri or crystallizing dish. The soil is then inoculated with a few drops of the mycelial slurry. For sand culture, pots filled with steam sterilized sand may be used in the following way. A small volume of sand is removed from the pot leaving a small depression sufficient to accommodate the seedling roots. The removed sand is mixed with mycelial macerate and replaced around the roots of the seedling as it is planted. Both these procedures result in young plants with more or less normal systems and typical ericoid mycorrhizas.

7.4.2   Arbutoid mycorrhizas

The fungal partners of arbutoid mycorrhizas are mostly basidiomycetes and the same or similar species to those that form sheathing mycorrhizas. One might expect that methods which are suitable for synthesizing sheathing mycorrhizas would also be successful for arbutoid mycorrhizas. Similar methods have in fact been used for the successful synthesis of mycorrhizas of *Arbutus menziesii* (Pacific Madrone) with *Pisolithus tinctorius*, *Thelephora terrestris*, *Corticium bicolor* and *Cenococcum geophilum*. These fungi were grown in Melin-Norkran's solution and the mycelium then macerated. This mycelial slurry was used to inoculate flasks of vermiculite/peat, moistened with Melin-Norkran's solution, in which surface-sterilized seeds of Pacific Madrone had been germinated. Mycorrhizas formed within 6 months.

8 Interaction between Fungal Symbiont and Host Plant

8.1 Sheathing mycorrhizas

There is much evidence that the host tree can derive several benefits from its mycorrhizal association. Those known are: (1) increased nutrient uptake; (2) increased water uptake and drought resistance; (3) resistance to certain root pathogens; (4) increased tolerance of toxins, temperature extremes and adverse pH. The fungal symbiont obtains its carbon and energy requirements from its tree partner, and when the symbiosis is established must benefit from access to nutrients with little or no competition from other microorganisms.

8.1.1 Nutrition

Striking growth responses of tree seedlings as a result of mycorrhizal inoculation leading to large increases in the dry weight of the mycorrhizal when compared with the non-mycorrhizal plants have been reported on many occasions. These growth responses are largely due to increased nutrient absorption and transport and may result in a higher nutrient concentration in mycorrhizal plants. For example, Hatch (1937) showed that infected pine seedlings grown in prairie soil absorbed 2–3 times more phosphorus, nitrogen and potassium per gram of roots than pine devoid of mycorrhizas. Growth responses are usually greatest in soils that are poor in mineral nutrients, particularly phosphate. It is not surprising that in most experiments growth responses are reduced or eliminated when mineral fertilizers are applied.

Although a wide range of both major and minor elements may be present in greater concentrations in mycorrhizal than in non-mycorrhizal plants, most studies on mineral uptake by sheathing mycorrhizas have been concerned with the absorption and fate of phosphorus, a mineral commonly deficient in natural soils. Indeed phosphorus deficiency is a typical symptom of mycorrhizal deficiency in the Pinaceae. It is probable that increased uptake of phosphorus by mycorrhizal roots leads to an increased rate of host metabolism and thence greater uptake of the other elements.

The uptake of nutrients, especially phosphate, by mycorrhizal roots appears to be controlled by several physiological processes including: (i) uptake by the fungal hyphae from the external soil solution; (ii) transfer from fungus via sheath and the intercellular hyphae (Hartig net) to the root cells; (iii) uptake by the root cells. Increased nutrient uptake can result either from increased rates of uptake, from an increase in the area absorbing, or from exploration of a greater soil volume. In some mycorrhizas, e.g. those of beech, the diameter of the root is increased as a result of the radial elongation of outer cortical cells. The sheath further increases the 'effective' diameter of the root and thus its surface area. In addition, there are usually hyphae or mycelial cords growing out into the soil; these effectively explore a much larger soil volume than non-mycorrhizal roots. However, some sheathing mycorrhizas, including those common in beech, are apparently smooth with few or no hyphae growing out into the soil.

One of the striking features of mycorrhizas is their sustained absorbing power that contrasts with the relatively transient absorbing power of uninfected roots. For example pine mycorrhizas are known to remain active for up to a year.

Studies on the uptake of phosphate from mineral solutions by excised beech mycorrhizas have shown that both the sheath and any radiating hyphae present strongly absorb mineral ions. From studies on beech and pine mycorrhizas it appears that as much as 80–90% of the absorbed phosphate accumulates in the sheath in the form of polyphosphate granules, similar to those in V–A fungi (see § 3.2 and § 8.2).

A high proportion of soil phosphorus may be present in organic compounds and it has frequently been suggested in the past without much evidence that sheathing mycorrhizal fungi may be able to enhance phosphate uptake by utilizing organic phosphates not available or less easily available to uninfected roots. The activity of acid phosphatase, able to degrade p-nitrophenyl phosphate, of excised beech mycorrhizas has been found to be up to 8 times that of uninfected roots. Although the activity of fungus and host were not separated, phosphatase activity throughout the sheath of beech mycorrhizas, as well as in six common symbionts of Douglas fir in culture, indicates the possible potential of the fungal partner to supply phosphate to its host, not only from the available inorganic pool but from organic compounds as well.

The process of nitrate and ammonium uptake by mycorrhizas from soils has been less thoroughly studied than that of phosphate. Excised beech mycorrhizas are known to absorb ammonium but not, to a significant extent, nitrate ions. However since many mycorrhizal fungi in pure culture produce nitrate reductase, and can therefore use nitrate as the sole source of nitrogen, it is likely that the mycorrhizas these form will be capable of utilizing nitrate from the soil. This may be of importance in some soils, but the majority of forest soils are acid and likely to contain little or no nitrate.

Mycorrhizas may benefit the plant under conditions of low soil water availability. Some mycorrhizal fungi can tolerate and grow in laboratory media with low water potentials. The common mycorrhizal fungus, *Cenococcun geophilum*, is very drought tolerant and has been particularly recommended for inoculating trees to be transplanted into dry areas.

As already mentioned (p. 8), the fungal partner absorbs inorganic nutrients from the soil and is the main source of these for the system. In contrast the photosynthesizing host tree is the primary source of carbohydrate for both partners, although there is evidence that the photosynthates may move not only from shoot to roots but also between root systems (i.e. from one plant to another) via mycorrhizal fungi. Experiments with excised beech mycorrhizas have shown that photosynthates are translocated to the root system where sucrose is hydrolysed by surface attached enzymes. From the products, the monosaccharides glucose and fructose are absorbed by the fungus into the sheath where they are rapidly converted to other products; glucose ending up as trehalose and glycogen and fructose as mannitol. Since these carbohydrates are now stored away in a form not readily available to the adjacent host tissues, a one way source to sink seems to be set up which ensures movement of carbon in the direction of the fungus.

8.1.2   Resistance to root pathogens

It has been demonstrated on several occasions that tree seedlings possessing sheathing mycorrhizas are more resistant to infection by soil-borne pathogens than seedlings with few or no mycorrhizas. For example, mycorrhizas formed by a variety of fungi on seedlings of both *Pinus taeda* and *P. echinata* conferred resistance to infections by zoospores and vegetative mycelium of *Phytophthora cinnamomi*. Non-mycorrhizal plants exposed to the pathogen showed reduced top growth, chlorosis, restricted root development and eventual death whereas the mycorrhizal plants grew normally. Findings such as these indicate that several mechanisms may operate. Possibilities include:

(*a*) secretion of antibiotics inhibitory to pathogens;

(*b*) the sheath acts as a physical barrier to penetration;

(*c*) surplus nutrients in the root are utilized, thus reducing the amount of nutrients available to pathogens; and

(*d*) the sheath supports a protective microbial rhizosphere population.

Well over 100 species of sheathing mycorrhizal fungi have now been shown to produce antibiotics. Furthermore diatretynes synthesized by the mycorrhizal fungus *Leucopaxillus cerealis* var. *pineina* were found to be inhibitory to *P. cinnamomi* in culture and probably responsible for the resistance of pines mycorrhizal with *L. cerealis* var. *pineina* to *P. cinnamomi*.

Evidence that the sheath of pine mycorrhizas can create a mechanical obstruction to pathogens attempting root penetration is based on extensive histological observation of pine mycorrhizas formed by non-antibiotic-producing fungal symbionts. As we have seen (Chap. 2), the sheaths of pine mycorrhizas are composed of tightly interwoven hyphae, often in well defined layers and usually completely covering the root meristem. When the sheaths are removed or incomplete, infection by *P. cinnamomi* readily occurs.

8.1.3   Increased tolerance of adverse conditions

Wide ecotypic differences both within and between fungal species ensure that there are fungi adapted to a wide range of environmental conditions (see Chap. 7). However, it is possible that within each habitat only a few fungi are capable of withstanding extreme conditions.

Examination of several coal spoils in the United States have shown that pine seedlings infected with *Pisolithus tinctorius* are capable of withstanding both low pH and high soil temperatures. More recently, at an arsenic toxic site in S.W. England with both high levels of soil arsenic and a pH of 2.9–4.0, fruit bodies of several mycorrhizal fungi were seen growing directly out of the steep face of the spoil. The importance of such observations for the reclamation of adverse sites will be discussed later (Chap. 9).

8.2   Vesicular-arbuscular mycorrhizas

There is now abundant evidence that in a great majority of cases V–A mycorrhizal infection of roots results in stimulation of plant growth. Since most natural untreated soils already contain V–A fungi, it is usual to sterilize, or partially sterilize, the soils being used in experiments to study the effects of inoculation with these fungi. This is generally done chemically with fumigants such as methyl bromide or by gamma radiation. It is accepted that these treatments themselves produce some changes in the soil likely to affect plant growth, for example changing the availability of certain nutrients, in addition to their effect on the microflora. However, many inoculation experiments have now been made both in pots and in the field using unsterilized soil, and a number of these have shown strong positive responses to inoculation.

A second problem in inoculation experiments with V–A fungi is in obtaining sufficient suitable inoculum. Since these fungi cannot be cultured axenically, roots of mycorrhizal plants with viable fungus are usually used as inoculum. With such inocula other organisms, particularly bacteria, will nearly always be introduced with the mycorrhizal fungi. Washings made from the V–A inocula will contain representative contaminating organisms, and so can be used to treat controls. Thus possible effects of these contaminants will occur in the controls and can be taken into account.

Growth responses in inoculation experiments have often been spectacular with the dry weight of inoculated plants being up to 50 times or even more that of uninoculated controls. As might be expected, such dramatic effects as these occur only in soils in which untreated plants grow very poorly. Usually this is because the soils contain very low levels of available phosphate and the growth response of inoculated plants is explained by much improved uptake of P by the mycorrhizas. A simple description of soil P and its use by plants would be that there are two major categories of P present in soil. The first of these, comprising orthophosphate ions, is readily utilizable by plants whether or not they are mycorrhizal. The second comprises a wide range of inorganic and organic compounds containing P, in equilibrium with the first and more or less unavailable to non-mycorrhizal plants. It is tempting to believe that the increased P uptake of mycorrhizal plants can be explained by access to forms of P in the second category not generally available to non-mycorrhizal plants. This possibility has probably been ruled out by a series of elegant experiments in which the soluble pool of phosphate in the soil has been uniformly labelled with ^{32}P. Mycorrhizal and non-mycorrhizal plants grown in these soils usually take up similar proportions of labelled and unlabelled P. If it were true that mycorrhizal plants had access to unlabelled insoluble inorganic or organic sources of phosphate in the soil they would then contain a smaller proportion of ^{32}P.

Studies of the uptake of nutrient ions by plant roots have shown that as specific ions are absorbd by roots a diffusion gradient towards them is created. Unless the nutrient is in good supply and diffusion is rapid, a zone immediately around the roots becomes progressively depleted and the rate of ion uptake decreases. This depletion and consequent decrease occurs quickly with P since phosphate ions diffuse relatively slowly. The external mycelium of V–A mycorrhizal fungi extends far beyond the depletion zone into undepleted soil. If it can absorb soluble phosphate and transport it to the root, the depletion zone will be effectively bridged and the supply of phosphate to the root increased. The technique, described in Chapter 6, in which most of the external mycelium is physically isolated from the root with which it is in connection, has confirmed that P is indeed conducted through the hyphae.

The rate of P flow or flux has been both calculated and measured and can be as high as 3.8×10^{-8} mol cm^{-2} s^{-1} in the rather large hyphae entering the root. This finding immediately posed the question, how could such rapid flow of P through the hyphae occur? Diffusion would not be nearly rapid enough, nor could unidirectional mass flow in the hyphae as a result of the plant's transpirational pull be the explanation. Continued mass flow in one direction would obviously soon empty the hyphae of all contents. The remaining possibility of cytoplasmic streaming is only feasible if the P could be contained in concentrated packets that could be moved along by an active streaming process. The polyphosphate granules seen by electron microscopy (see § 3.2) and identified histochemically could provide just such concentrated phosphate packets. It now seems very likely that it is the movement of these granules with their associated small vacuoles by active cytoplasmic streaming that accounts for the rapid flow of P through external hyphae to the root. Large responses to V–A mycorrhizal infection usually occur in soils deficient in available P. Experiments with some plants, in particular the tree sweetgum (*Liquidamber styraciflua*), have shown very good responses to V–A inoculation even when high rates of an N-P-K fertilizer have been added to the soil. Comparable results have been obtained in experiments with sheathing mycorrhizas and tree seedlings. Results such as these suggest that the availability of P in the soil is only part of the story of responses to mycorrhizal inoculation.

The enhanced supply of P in plants having V–A mycorrhizas may have a special significance in leguminous plants. In P-deficient soils legumes often nodulate very poorly or not at all unless they become mycorrhizal. The improved supply of P following infection by mycorrhizal fungi causes a strong stimulation of nodulation, which in turn results in an increased supply of nitrogen that contributes further to growth stimulation.

It is not really surprising that if a plant grows more vigourously, following V–A fungal infection, due to increased P uptake, it should also be able to take up larger quantities of other nutrients. When enhanced uptake of other nutrients occurs it is not always easy to decide whether it is due to the larger root system of the mycorrhizal plant or to a direct effect of the mycorrhizal state. Some experimental results do suggest a direct effect. For example, experiments with a steam-sterilized potassium-deficient soil showed that inoculated plants of *Griselinia littoralis*, an evergreen New Zealand shrub, grew better and took up more potassium (K) than uninoculated ones. When K was added to the soil, inoculation did not produce a response, clearly showing that in this case the mycorrhiza was enhancing K uptake and hence growth. A reason why improved plant growth resulting from increased K uptake following inoculation is likely to be much rarer than an effect on P is because K has a diffusion rate 10–20 times that of P. Depletion zones for K cannot therefore easily develop around roots. A further example where another nutrient was involved was the much improved growth of peach seedlings following inoculation with V–A fungi. Uninoculated seedlings were stunted and showed severe symptoms of zinc deficiency which largely disappeared following inoculation.

Various other beneficial effects besides improved nutrient uptake have been associated with V–A mycorrhizas. These include improvement in water transport in the plant and increased resistance to root but not to shoot pathogens. Another way in which V–A fungi may be beneficial is by improving aggregation of soil particles. This has been observed to occur in a sand dune soil where the sand particles adhered strongly to *Glomus* hyphae and bean roots.

As we have seen in § 3.4, several genera and species of V–A fungi have been recognized. It has been established that individual species usually have a wide host range and, as a corollary, that several different species of V–A fungi may be able to infect one host species. In experiments in which the effects of different species of V–A fungi on a host were compared B. Mosse found large differences. Growth of onions was increased from 2–15 times and that of the tropical grass *Paspalum notatum* from 2–10 times, depending on the fungal species used. These results were obtained in one soil. With different soils the relative stimulation produced by mycorrhizal species varied and was also affected by whether or not lime was added to the soil. It also seems that the ability of a particular V–A fungus to infect a plant may be pH dependent. The significance of these findings is that, although specificity in the normal sense does not occur, there are important differences in the effectiveness of different fungus/host combinations. Further, different V–A species vary in their ability to establish infections in different soils. It follows from the finding that V–A fungi differ in their effects on growth that naturally occurring species or strains may not always be the most effective. If more effective strains could replace less desirable indigenous strains in a particular environment, benefit should result. Such replacement can sometimes be achieved by inoculation.

8.3   Orchid mycorrhizas

Orchids depend for at least a part of their life cycle on their fungal symbiont to provide them with their carbon and energy supply (see Chap. 5). So, in their natural environment at least, orchids are obligately mycorrhizal and questions about growth stimulation resulting from infection, or other possible benefits, may not seem relevant. However, it is possible that the symbiosis is beneficial to the orchid in other ways, for example in its mineral nutrition or by providing growth substances or vitamins.

8.3.1   Nutrition

The nutrition of orchid mycorrhizal fungi differs considerably between species. It appears to be related to the ecology of the fungi with more specialized slowly-growing symbionts, such as *Tulasnella calospora* (see § 5.3), being nutritionally exacting, while less specialized species such as

Ceratobasidium cornigerum are faster growing and not nutritionally exacting. A simple mineral salt solution containing nitrate nitrogen and glucose supports good growth of *C. cornigerum* and some other *Rhizoctonias*, but isolates of *T. calospora* do not grow on it at all. They require for good growth a suitable organic source of nitrogen, the amino acids asparagine and glycine are particularly good, and the vitamins thiamine and p-amino benzoic acid. The need of these nutritionally-exacting fungi for amino acid nitrogen and vitamins may be satisfied from external sources in the soil or, once penetration has occurred, from the host orchid tissues.

It is also likely that orchid mycorrhizal fungi play a part in the mineral nutrition of both seedling and adult orchids in a manner comparable with other mycorrhizal systems. This is supported by the observation that P translocation is known to occur in the hyphae of an orchid mycorrhizal fungus.

It is in their carbohydrate nutrition that orchid mycorrhizal fungi differ from almost all others. Those species that have been tested grow well with cellulose as a carbon substrate and a few can utilize lignin. More important, when glucose and cellulose have been compared as carbon sources for symbiotic orchid seedlings, cellulose has proved superior in promoting protocorm growth. Particularly with the more aggressive fungi, there is a greater likelihood of the fungus becoming parasitic on the orchid when supplied with glucose as compared to cellulose. Asymbiotic protocorms of at least some orchids accumulate much starch, whereas in symbiotic protocorms starch synthesis appears to be suppressed while growth is accelerated.

Some orchid mycorrhizal fungi can grow and utilize cellulose in unsterile soil in competition with other fungi and experiments with ^{14}C have shown that absorbed carbon is translocated through hyphae into plant tissue. The nature of the carbohydrate that passes from the fungus to the plant appears to be the fungal sugar, trehalose. The fact that some orchids can utilize trehalose very efficiently makes it likely that this carbohydrate is transferred directly to the orchid tissues, where it is converted to sucrose.

8.4   Ericaceous mycorrhizas

Since ericaceous plants usually grow naturally in nutrient-poor soil (see § 4.1), the possession of mycorrhizas might be expected to be an advantage. There have long been suspicions that the N nutrition of at least some ericaceous plants is improved by their mycorrhizal association. One theory, now discredited, was that the mycorrhizal fungus fixed atmospheric N. As we shall see, more recent experimental work shows the probable role of some ericaceous mycorrhizal fungi in the N nutrition of their hosts.

8.4.1   Nutrition

It is only recently that we have had much knowledge of the physiology of the mycorrhizal association. There is detailed experimental evidence that ericoid mycorrhizal infection stimulates the growth of seedlings. This may be at least partly due to increased uptake and translocation of P, as in sheathing and V–A mycorrhizas. This possibility is supported by the observation that experimentally infected seedlings of *Calluna vulgaris* and *Vaccinium macrocarpon* contain significantly more P than non-mycorrhizal controls. Also, the endophyte of *C. vulgaris* shows considerable phosphatase activity in culture. Such activity, if it occurred when the fungus was mycorrhizal, would have the potential to increase the P supply to the plant.

Peaty heathland soils, where ericaceous plants are often dominant, are usually characterized not only by low levels of available P, but also by low levels of available N. Much of the N that is present is complexed in organic compounds. Also, particularly at high altitudes, mineralization of N in these soils tends to be very slow. Experimental evidence that the ericoid type of mycorrhizal infection enhances N uptake from soil is therefore particularly significant. The increased uptake of inorganic N by mycorrhizal seedlings of *Vaccinium macrocarpon* from solutions containing low concentrations of ammonium sulphate may help to explain this. However, the ability of the mycorrhizal fungus to utilize organic sources of N (see § 6.4) could be even more important in this context. Recent experiments have shown that young mycorrhizal plants of *V. macrocarpon* can utilize amino acids as a nitrogen source, an ability not so well developed in non-mycorrhizal plants. The conclusion from our present knowledge of the mineral nutrition of ericoid mycorrhizas is that the association is adapted to benefit plant growth in nutrient-poor soils; the typical habitat of ericoid plants. It is tempting to speculate that other mycorrhizal associations, particularly sheathing mycorrhizas of trees growing in deep peats, may be beneficial partly because they improve nitrogen uptake.

Little is known of the physiology of the arbutoid mycorrhizas of *Arbutus* and *Arctostaphylos*. The recent discovery of polyphosphate granules in hyphae of the mycorrhizal fungus of *Arbutus unedo* tends to confirm the suggestion that *Arbutus* mycorrhizas may resemble sheathing mycorrhizas both functionally and physiologically.

Studies of the arbutoid mycorrhiza of the epiparasite *Monotropa hypopitys* give stong evidence that nutrients, including phosphorous, are transferred from trees, via shared mycorrhizal fungi, to *Monotropa*. When ^{32}P-labelled orthophosphate was injected into the phloem of mature conifers it was later detected in *Monotropa* plants up to 2 metres away. However, little or no ^{32}P was found in other nearby plant species. Although it thus seems that the fungal symbiont can be regarded as a nutrient bridge, it is not clear what proportion of its nutrients *Monotropa* gets this way. At least part of its nutrient supply is likely to be from the soil through the masses of mycorrhizal fungal hyphae that envelope its roots. There are two periods in the year when *Monotropa* needs to mobilize large amounts of nutrients (§ 4.3.2). It may be at these times that the reserves of shared mycorrhizal trees are of particular importance.

All the host plants of ericoid mycorrhizal fungi are green and autotrophic with respect to carbon compounds. Carbon dioxide labelled with ^{14}C has been used to incorporate ^{14}C in the photosynthate in shoots of mycorrhizal *Vaccinium* seedlings. This label has been followed into the roots and thence the external mycelium of the fungal endophyte, where it accumulates in the fungal carbohydrates mannitol and trehalose. This observation, together with the fact that the endophyte has only limited cellulolytic ability, indicates that, like sheathing mycorrhizal fungi, ericaceous mycorrhizal fungi depend on their host plant for at least most of their carbohydrate supply.

The ericoid mycorrhizal relationship thus seems to be mutualistic with the host plant receiving mineral nutrients from its fungal partner, which in turn benefits by the carbon compounds it obtains from its host. In sharp contrast, the achlorophyllous *Monotropa* relies on its fungal partner for both its mineral and carbon nutrition. There is now conclusive evidence that carbon compounds move from forest trees to *Monotropa* plants via the shared mycorrhizal fungus. Although ericaceous mycorrhizas are very diverse, it does seem that the morphologically distinct ericoid mycorrhizas and the arbutoid mycorrhizas of *Arbutus* and *Arctostaphylos* are physiologically very similar and therefore comparable to the sheathing mycorrhizas of trees. In contrast, *Monotropa* mycorrhizas, although morphologically similar, to sheathing and other arbutoid mycorrhizas, are much more similar functionally to those of the heterotrophic orchids.

Table 8–1 summarizes the more important evidence and hypotheses of the nutrient inter-relationship of mycorrhizas.

Table 8–1   The probable main nutrient fluxes between soil, fungus and plant.

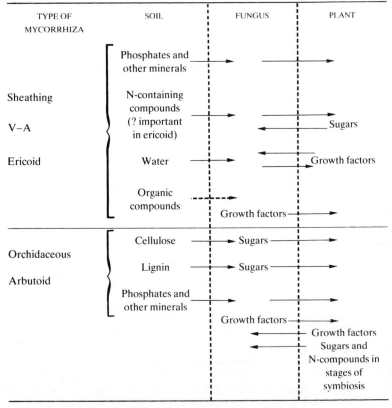

Dotted lines indicate less important or hypothesized fluxes.

9 Applications of Mycorrhizas

9.1 Introduction

From the preceding chapters it will be realized that mycorrhizas probably always affect a plant's growth and sometimes its health, and are usually beneficial and sometimes essential. Since the importance of mycorrhizas to the plant's well being was first suspected, there has been an increasing interest in the possibilities of maximizing their beneficial effects.

In situations where mycorrhizas would not otherwise occur, any action that results in their developing is likely to improve plant growth or even in extreme cases to make it possible when otherwise it would not have been. More often there is a situation in which, with normal methods of planting and cultivation, mycorrhizal associations do develop but they may not be with fungi that give the greatest possible benefit. Also, there may be a considerable delay before mycorrhizas become established. If a more effective mycorrhizal fungus can be successfully introduced and if it can be ensured that the association develops early in the plant's growth, then benefit is likely to result.

Most attempts to introduce mycorrhizal fungi have been by using selected fungi as inocula. These fungi are used either as pure cultures or, in the case of V–A fungi, as mycelium and spores developing in association with live roots. In some situations inocula of sheathing mycorrhizas have consisted of soil, generally containing live propagules of several potential mycorrhizal fungi. There is also the possibility that by suitable methods of cultivation beneficial mycorrhizal associations with fungi already present may be encouraged.

Much of the present interest in mycorrhizas is concerned with the possibilities of applying our knowledge to our advantage. As the need to conserve energy and reserves of natural mineral fertilizers increases, so work on the application of mycorrhizas is likely to intensify. The optimum utilization of mycorrhizal associations could make an important contribution to the production of food and other plant products, particularly in third world countries. Hopes for the applications of mycorrhizas are encouraged by the increasing research and successes in the biological control of pests and diseases.

A special application of mycorrhizas, of great potential importance, is in the replanting of land made derelict, usually by industry. Such land is often very poor in plant nutrients and may also be, in varying degrees, phytotoxic. There are several objectives of replanting such land. These include restoration and improvement of the environment for its amenity value, stabilization of the soil and the production of crops. The presence of mycorrhizal fungi adapted to the stressed conditions so frequent in such land is often essential to success in replanting.

9.1.1 Plant food crops

Considering the complexities of the interacting factors in the root zone of plants growing in natural (unsterilized) soil, it is not altogether surprising that the outcome of inoculation experiments with mycorrhizal fungi is not predictable and may not always be reproducible. There have now been sufficient instances of significant and often quite large responses in the growth and yield of crops to inoculation with V–A fungi to encourage more intensive research.

As shown in § 8.2 in inoculation experiments in sterilized soils and in which control plants grow poorly, responses have frequently been very large. In the field or in commercial glasshouses such large responses cannot be expected. Two reasons for this are that natural V–A infections are likely to occur and that in normal practice fertilizers will usually be used to ensure reasonable yields. Inoculation with V–A fungi becomes of practical importance if the outcome is more or less predictable and if it is cost-effective in terms of increased yield relative to the expense of inoculation.

The results of many field experiments on inoculation of different crops with V–A fungi have now been published. In several of these experiments moderate to large growth increases resulted from V–A inoculation, even though the soil was not sterilized in any way. Table 9–1 summarizes some of these results in which unsterile soil was used.

A criticism of some of these experiments is that inoculated seedlings were transplanted into the field; when normally the crops would be sown in the field directly. However, in other experiments infective soil or soil plus roots was used as inoculum at the time of sowing. Such experiments show the potential of V–A mycorrhizal inoculation, but leave some questions unanswered. In particular, how sufficient inoculum could be produced to inoculate plants on a large scale, whether inoculation one year would also benefit crops in subsequent years and whether an economic return could be expected with field crops such as cereals.

A field situation somewhat comparable with pot experiments using sterilized soil may arise when soil fumigants are used. Fumigants such as methyl bromide are used to control nematodes and other soil-borne pests of citrus. It has been noticed on several occasions that citrus seedlings planted in fumigated soil became stunted and chlorotic. This turned out to be due chiefly to phosphorus deficiency resulting from a complete absence of V–A fungi. Inoculation of the seedlings with V–A fungi gave normal growth in the fumigated soils. As the practice of soil fumigation increases, instances where inoculation with V–A fungi is the best way of restoring good growth and subsequent crop yield will also increase.

9.1.2   Fungal food crops

An interesting and important 'by-product' of many sheathing mycorrhizal fungi is their edible fruit body. Amongst the most highly esteemed of these is the truffle; the underground fruit body of species of *Tuber*. The best edible species, *T. melanosporum*, occurs naturally in very limited areas in Europe, mostly in France in association with oaks, hazel, sweet chestnut and occasionally other trees. Its occurrence is sporadic and it is difficult to collect. Progress has been made in the cultivation of truffles in France by inoculating seedlings of oak and hazel with minced up fruit body tissue. With favourable

Table 9–1 Responses of various crops to inoculation with V–A fungi.

Crop	V–A species	Growth increase	Country	Author
Maize	*Glomus mosseae*	× 2.3	Pakistan	Khan (1972)
Wheat	*Glomus mosseae*	× 3	Pakistan	Khan (1975)
Barley	*Glomus mosseae*	× 4	Pakistan	Saif and Khan (1977)
Barley	*Glomus caledonius*	× 1.3	United Kingdom	Owusu-Bennoah and Mosse (1979)
Onion	*Glomus caledonius*	× 6	United Kingdom	Owusu-Bennoah and Mosse (1979)
Lucerne	*Glomus caledonius*	× 4	United Kingdom	Owusu-Bennoah and Mosse (1979)
Maize	*Glomus mosseae*	× 1.3	Nigeria	Islam (1977)
Cowpea	*Glomus mosseae*	× 1.3	Nigeria	Islam (1977)
Cotton	*Glomus fasciculatus*	× 2.2	India	Bagyaraj and Manjunath (1980)
Cowpea	*Glomus fasciculatus*	× 1.5	India	Bagyaraj and Manjunath (1980)
Finger millet	*Glomus fasciculatus*	× 3.2	India	Bagyaraj and Manjunath (1980)

soil and climatic conditions, the first truffles can be produced by these inoculated mycorrhizal trees in as little as 3½ years. About 65 000 inoculated tree seedlings are now being planted annually.

The food value of other mycorrhizal fungi such as species of *Boletus*, *Leccinum* and *Lactarius* should be an important criterion in selecting fungi for inoculating forest tree seedlings. In France the application of fertilizers to existing stands of beech and oak is being practiced to encourage the fruiting of sufficient quantities of edible mycorrhizal fungi for selling in local markets.

9.2   Forestry and amenity trees

The practical importance of mycorrhizas in forestry was first noticed when plantations of exotic (i.e non-indigenous) pines in different parts of the world, especially tropical countries, almost invariably failed until suitable fungi were introduced in one way or another. Originally they were introduced in soil or mycorrhizal seedlings from areas where the pines thrived, or broken up sporophores of mycorrhizal fungi were used as inoculum. Such simple methods as these were usually successful in causing large responses in tree growth. Planting trees on mine spoils and similar sites has some analogies to planting exotic pines as these sites may be completely devoid of potential mycorrhizal fungi.

There is now increasing interest in the potential benefits of inoculating tree seedlings with cultures of selected mycorrhizal fungi that may be of greater value to the tree than natural infection by fungal species already present in the nursery and planting site. Such inoculation with carefully selected fungi may be of particular value in the establishment of tree species such as Sitka spruce on difficult soils, for example deep acid peats.

Pure cultures of selected mycorrhizal fungi, often grown on a mixture of peat and vermiculite plus a nutrient solution, may be applied directly to the tree nursery beds or incorporated in the compost in which containerized tree seedlings are being grown. The latter is a particularly appropriate method for amenity trees. For large scale inoculation of trees bulk inoculum is produced using industrial scale fermenters.

Figure 9–1 shows the mean dry weight of shoot and root systems of Sitka spruce seedlings grown for 6 months in fumigated forest soil with either no mycorrhizal inoculum or mycelial inoculum of seven different fungi. The five fungi that formed mycorrhizas in this experiment all produced large growth increases. *Pisolithus tinctorius* failed to form mycorrhizas and did not stimulate growth. It is interesting that the two species of *Laccaria* caused significantly different responses as did the two different isolates of *Thelephora terrestris*. In other experiments *Paxillus involutus* was able successfully to colonize Sitka spruce seedlings growing in unsterile soils and, like *Laccaria* sp., it significantly increased its host's growth.

Although it is true that the majority of forest trees planted have sheathing mycorrhizas, there are also important species, both of forest and amenity trees, particulary in the U.S.A. and also in the tropics, that have V–A mycorrhizas. These include sweetgum (*Liquidamber styraciflua*), the tulip tree

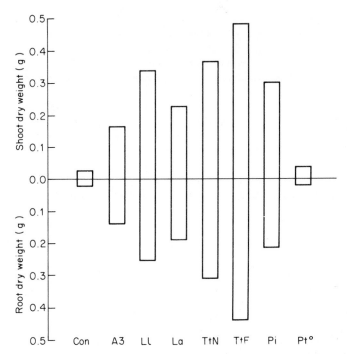

Fig. 9–1   Shoot and root dry weights of Sitka spruce seedlings inoculated with different mycorrhizal fungi. **Con**, control (not inoculated); **A3**, unknown basidiomycete; **Ll**, *Laccaria laccata*; **La**, *Laccaria amythestea*; **TtN**, *Thelephora terrestris* (nursery isolate); **TtF**, *Thelephora terrestris* (forest isolate); **Pi**, *Paxillus involutus*; **Pt**, *Pisolithus tinctorius* (no mycorrhizas formed). (By courtesy of Dr J.M. Holden.)

(*Liriodendron tulipifera*), maples (*Acer* spp.), alders (*Alnus* spp.) and poplars (*Populus* spp.). The last three genera may have either sheathing or V–A mycorrhizas and sometimes both on the same species or even on the same individual. Research in America on the sweetgum seems to show that it has an obligate dependence on V–A mycorrhizal infection. Even with apparently adequate mineral fertilization, non-mycorrhizal plants grow very poorly or even fail. In one experiment with fumigated soil receiving adequate amounts of phosphorus and other essential elements inoculation with *Glomus mosseae* and a natural soil inoculum increased plant dry weight 34 and 56 times respectively.

9.3   Land reclamation

Large areas of waste land are created by mining and other industrial operations. These areas are usually more or less unfavourable for the re-establishment of plants. The nature of the wastes produced varies widely and includes such diverse materials as colliery spoil, sand and gravel waste, silica, china clay, metal mine spoils, chemical wastes, domestic and sewage refuse and agricultural wastes. Each has its peculiar problems including extremes of pH, infertility, toxicity, erosion and instability. It is not surprising that restoration can be very costly, often involving such steps as reshaping and re-grading the waste, correction of nutrient deficiencies and correction of pH. Even after these have been done, up to 90% of plants may fail to establish.

Research in Britain and the U.S.A. has shown that a common reason for this is failure of plants to develop their mycorrhizas, whether these be sheathing or V–A. Conversely, plants that do succeed in naturally colonizing mine spoils and similar materials are usually mycorrhizal. However, this may not always be true. For example, in disturbed areas of the Red Desert in Illinois the predominant colonizers are non-mycorrhizal Old World species of the Chenopodiaceae, while in undisturbed areas mycorrhizal New World species predominate. It is argued that plants colonizing the disturbed areas are species that adopt a ruderal strategy, and that being mycorrhizal would be of no value to them in sites that are usually devoid of mycorrhizal inocula.

Growth experiments in mine spoils have shown that plants with mycorrhizas usually grow better than non-mycorrhizal ones. This is not surprising since phosphate is usually predominant amongst the mineral nutrients that are deficient in such material. Red maple seedlings inoculated with *Gigaspora gigantea* and grown in anthracite waste to which bonemeal had been added had, after 70 days, more than five times the dry weight of seedlings that were not inoculated. Similarly recent research in America has shown that the sheathing mycorrhizal fungus *Pisolithus tinctorius* significantly increases survival and growth of pines planted on a range of adverse industrial waste sites. In Britain, the related *Scleroderma citrinum* as well as *Paxillus involutus* are common symbionts of trees on coal tips and inoculation experiments with these fungi are in progress.

Since nitrogen is often amongst the minerals deficient in industrial wastes, it is not surprising that there may be advantages in planting them with legumes inoculated with the appropriate species of the nitrogen-fixing bacterium *Rhizobium*. Still greater advantages can result from using legumes inoculated with both *Rhizobium* and V–A mycorrhizal fungi (see § 8.2), or plants such as *Alnus* and *Elaeagnus* that have a nitrogen-fixing symbiosis with actinomycetes and also form V–A mycorrhizas.

References

BAGYARAJ, D.J. and MANJUNATH, A. (1980). Response of crop plants to V–A mycorrhizal inoculation in an unsterile Indian soil. *New Phytologist*, **85**, 33–6.

BERNARD, N. (1909). L'évolution dans la symbiose. Les Orchidées et leur champignons commensaux. *Annales des Sciences Naturelles Series,* **9**. 1–196.

DEVERALL, B.J. (1981). *Fungal Parasitism*, second edition. Studies in Biology No. 17. Edward Arnold, London.

GERDEMANN, J.W. (1955). Relation of a large soil-borne spore to phycomycetous mycorrhizal infections. *Mycologia,* **47**, 619–32.

GERDEMANN, J.W. and TRAPPE, J.M. (1974). The Endogonaceae in the Pacific Northwest. *Mycolgia Memoirs,* **5**, 1–76.

HATCH, A.B. (1937). The physical basis of mycototrophy in the genus *Pinus*. *Black Rock Forest Bulletin,* **6**, 1–168.

ISLAM, R. (1976). Effect of several Endogone spore types on the yield of *Vigna uniguiculata*. International Institute of Tropical Agriculture, Ibadan, Nigeria. IITA Internal Report.

KHAN, A.G. (1972). The effect of vesicular-arbuscular mycorrhizal associations on growth of cereals I. Effects on maize growth. *New Phytologist,* **71**, 613–19.

KHAN, A.G. (1975). The effect of V–A mycorrhizal associations on growth of cereals II. Effects on wheat growth. *Annals of Applied Biology,* **80**, 27–36.

LEWIS, D.H. (1973). Concepts in fungal nutrition and the origin of biotrophy. *Biological Review,* **48**, 261–78.

MASON, P.A. (1975). The genetics of mycorrhizal associations between *Amanita muscaria* and *Betula verrucosa*. In: *The Development and Function of Roots*, J.G. Torrey and D.T. Clarkson (eds), pp. 567–74. Academic Press, London.

MASON, P.A. (1980). Aseptic synthesis of sheathing (ecto-) mycorrhizas. In: *Tissue Culture Methods for Plant Pathologists*, D.S. Ingram and J.P. Helgeson (eds), pp. 173–8. Blackwell Scientific Publications, Oxford.

MARX, D.H., HATCH, A.B. and MENDICINO, J.F. (1977). High soil fertility decreases sucrose content and susceptibility of loblolly pine roots to infection by *Pisolithus tinctorius*. *Canadian Journal of Botany,* **55**, 1569–74.

MOSSE, B. and PHILLIPS, J.M. (1971). The influence of phosphate and other nutrients on the development of vesicular-arbuscular mycorrhiza in culture. *Journal of General Microbiology,* **69**, 157–66.

OWUSU-BENNOAH, E. and MOSSE, B. (1979). Plant growth responses to vesicular-arbuscular mycorrhiza XI. Field inoculation responses in barley, lucerne and onion. *New Phytologist,* **83**, 671–9.

PATE, J.S. and GUNNING, B.E.S. (1972). Transfer cells. *Annual Review of Plant Physiology,* **23**.

SAIF, S.R. and KHAN, A.G. (1977). The effect of vesicular-arbuscular mycorrhizal associations on growth of cereals III. Effects on barley growth. *Plant and Soil,* **47**, 17–26.

SCOTT, G.D. (1969). *Plant Symbiosis*. Studies in Biology No. 16. Edward Arnold, London.

SINGER, R. (1975). *The Agaricales in Modern Taxonomy*. J. Cramer, Vaduz, Germany.

Index (**Bold** numbers refer to illustrations)